COMMEI
ARITHMETIC

The object of this book is to explain simply and clearly
the basic principles of arithmetic, and to enable the
reader to acquire facility in their application to such
problems as he or she is likely to encounter in business
and everyday life. It covers the elementary examination
syllabuses of the Royal Society of Arts, the London
Chamber of Commerce, and similar examining bodies.

TEACH YOURSELF BOOKS

The book covers fully all the usual aspects of arithmetic with some excellent chapters at the end on Foreign Exchange, Rates and Taxes, Bills of Exchange, etc. The student who conscientiously works through the well-chosen examples cannot fail to get a good grounding in the subject.

Scottish Educational Journal

COMMERCIAL ARITHMETIC

J. H. Harvey

TEACH YOURSELF BOOKS

Hodder and Stoughton

First printed 1949
Second edition 1970
Third edition 1972
Fourth edition 1977
Fifth edition 1982
Fourth impression 1988

British Library Cataloguing in Publication Data
Harvey, J. H.
   Commercial arithmetic. – 5th ed. – (Teach Yourself
series)
   1. Business mathematics
   I. Title
   513′.93   HF5691

ISBN 0 340 28269 X

Printed in Great Britain for
Hodder and Stoughton Educational,
a division of Hodder and Stoughton Ltd,
Mill Road, Dunton Green, Sevenoaks, Kent,
by Richard Clay Ltd, Bungay, Suffolk

Introduction

The purpose of this volume is to set out in clear and simple language the elementary principles of Arithmetic and to enable the student to acquire facility in their application to such problems as he, or she, is likely to encounter in business and everyday life. The problems given in the exercises for practice are, in general, of a thoroughly practical nature. The text is presented as simply as is consistent with a clear understanding of the principle being explained. The work is intended primarily for the private student, and for those engaged in business who may be interested in this subject but may be unable to attend classes. Teachers may find it useful as an aid in the preparation of notes of lessons, and it could, of course, be used as a class textbook. The book covers the elementary examination syllabuses of the Royal Society of Arts, the London Chamber of Commerce, the Business Education Council, and similar examining bodies, and it could therefore be used with advantage by those students who contemplate sitting for such examinations.

The work is modelled on the lines of the author's book, *The Arithmetic of Commerce*, and he acknowledges his indebtedness to the Gregg Publishing Company for permission to use certain diagrams and tables from that book.

Note to the 1982 Edition

Mr Harvey asked me if I would revise his books after his death and this edition is my attempt to do that. Should any reader have any suggestions for improvement, I would be delighted to consider them for inclusion in subsequent editions.

Although it is very difficult to bring a book on this subject truly up to date because of the current international financial instability with rapidly changing commodity prices and exchange rates, I have tried to make wages and prices as realistic as possible, and the section on taxation has been completely revised. Certain other textual changes have also been made to bring this edition up to date.

In any case, what is really important are the methods used in this book and if these are learned and understood thoroughly, then the figures used in the computations are immaterial.

A word or two on the use of electronic calculators is worthwhile here. These cheap, reliable and accurate machines are an extremely useful tool in the world of commerce and arithmetic and have brought about a small revolution in this field. However, the student must learn to do the basics unassisted in longhand and thus to be entirely self-sufficient. Not only is this a wise precaution against mechanical or electrical failures, but in the Elementary stages, neither calculators nor logarithm tables are permitted in examinations. A calculator should only be used to *check* results, not to arrive at them.

R H Statham

Contents

Hints to the Student

1. Follow carefully the working of each example before working an exercise.

2. Read each question *carefully* before working.

3. When working an exercise, set out each problem *clearly* and *neatly*.

4. *Never* look at the answer before you have attempted a problem. The answer is given as a check on your working, and to look at it beforehand may tend to destroy your confidence in your own ability.

5. When short methods are given, study them *very carefully*, and make quite sure that you understand them thoroughly before using them. It is preferable to use a long method which you understand than a short method of which you are not quite sure.

6. Whenever possible *check* your answer by working the question in another way.

First Four Rules—Simple Accounts

1. Addition

Addition is the process of finding the sum of two or more given numbers.

A great deal of the practical work of commercial calculations consists of the addition of columns of figures.

Speed and accuracy in addition can be attained only by practice.

The symbol + (plus) is used to denote addition.

The symbol = (equal or equals) denotes equality.

The result of addition is called the *sum* or *amount*.

Thus, if we wish to indicate that 7 plus 8 equals, or is equal to, 15, we write $7 + 8 = 15$.

We should educate the eye to take in at a glance the sum of two figures and, as far as possible, exclude unnecessary words from our minds when adding.

Taking the above example, we should *not* say, either orally or mentally, 7 *and* 8 make 15, but we should look at 7 and 8 and mentally register 15.

Ex. 1 $4718 + 396 + 1457 + 16\,954 + 18.$

Place these figures in a column, being very careful in correctly aligning them in their place values, *i.e.* units under units, tens under tens, hundreds under hundreds, etc.

$$
\begin{array}{r}
4\,718 \\
396 \\
1\,457 \\
16\,954 \\
18 \\
\hline
23\,543 \\
\end{array}
$$

The mental steps are as follows:

(1) 8, 12, 19, 25, 33. Write down 3. Carry 3.
(2) 4, 14, 24.   ,,   4.   ,,   2.
(3) 11, 15, 25.   ,,   5.   ,,   2.
(4) 18, 19, 23.   ,,   23.

The addition can now be checked by adding each column from the top to the bottom. Some students may find it easier to add up the first column, down the second, up the third, down the fourth, and so on. You can find the method which best suits you by trying both.

The following should be noted:

In (2) there was a jump from 4 to 14 and from 14 to 24, the two 5's and the 9 and 1 immediately registering 10.

In (3) 7 and 3 immediately registered 10.

Consider the following:

Ex. 2   Add horizontally into the blank spaces on the right. Then add the three vertical columns.

£	£	£
16·79	36·41	———
19·36	9·38	———
27·47	17·59	———
5·94	43·27	———
£	£	£

Solution:

£	£	£
16·79	36·41	53·20
19·36	9·38	28·74
27·47	17·59	45·06
5·94	43·27	49·21
£69·56	£106·65	£176·21

The mental steps follow the same pattern as in Ex. 1, but allow for correct placing of the decimal point between pounds and new pence.

It should be observed that the total of the third column is also the total of the totals of the first and second columns. It therefore acts as a check figure, and if the two agree it may be assumed that the additions are correct. It is, of course, possible to make a compensating error, that is, exactly the same error vertically and horizontally, but this is very unlikely to happen. The student who wishes to attain skill in addition may, therefore, make out his own tots with as many lines and as many columns as he wishes, using the grand total as a check on his work.

2. Subtraction

If we subtract 7 from 11 we get an answer of 4. 4 is the difference between 7 and 11. If we added 4 to 7 we should get 11. There are, therefore, two methods by which subtraction may be performed.

Ex. 3   6842 − 5719.

$$6842$$
$$5719$$
$$\overline{1123}$$

Mental steps:

1st method.

    (1) 9 from 2 is impossible, so we take 1 from the next higher denomination, making 12. 9 from 12 = **3**

    (2) We have taken 1 from 4, leaving 3. 1 from 3 = **2**

    (3) 7 from 8 = **1**

    (4) 5 from 6 = **1**

2nd method.

    (1) 9 and 3 = 12. Write down **3**

    (2) 1 and 2 = 3.   ,,   **2**

    (3) 7 and 1 = 8.   ,,   **1**

    (4) 5 and 1 = 6.   ,,   **1**

4 *Commercial Arithmetic*

3. Complementary Addition

This is the 2nd method of subtraction illustrated in Example 3.

Complementary numbers are numbers which, when added together, give a required total.

In Example 3, 1123 and 5719 are complementary numbers to the total 6842. 1123 is the complement, difference, or remainder.

Ex. 4   Total and balance the following:

£		£
547·19		27·13
856·36		1442·47
19·58		215·89
326·29	Balance	
£		£

The *Balance* is the amount necessary to make the two columns add up to the same amount. It is the difference between the greater amount and the lesser amount. It can be found by either of the above methods of subtraction.

1st method. Ordinary subtraction.

    (*a*) Total the first column.
    (*b*) Total the second column and write the total on a separate piece of paper.
    (*c*) Subtract the total of the second column from the total of the first column.
    (*d*) Insert the difference (*i.e.* the balance) in the second column.
    (*e*) Add up the second column, the total of which should be the same as the total of the first column.

2nd method. Complementary addition.

    (*a*) Total the first column.
    (*b*) Place the same total in the second column.
    (*c*) Obtain the balance by complementary addition.

It will be obvious that the second method is the better. We will now give the working of Example 4, using the method of complementary addition.

£		£
547·19		27·13
856·36		1442·47
19·58		215·89
326·29	Balance	63·93
£1749·42		£1749·42

Steps:
  (1) Total 1st column, £1749·42.
  (2) Write this down in both columns.
  (3) Balance by complementary addition method.

Mental steps:
  (*a*) Pence. Units: 10, 19.
        Write down 3, = 22. Carry 2.
              Tens: 10, 15.
        Write down 9, = 24. Carry £2.
  (*b*) Pounds. 7, 9, 16.
        Write down 3, = 19. Carry 1.
          2, 6, 8.
        Write down 6, = 14.

4. Simple Accounts

An *Account* is a record of business transactions. In business, accounts are kept in a book called the Ledger, which is usually ruled as in Example 5. The left-hand side is known as the *Debit* side, and the right-hand side is known as the *Credit* side. Entries on the left-hand side are debit entries, and entries on the right-hand side are credit entries.

A *Debtor* is a person who owes money.

A *Creditor* is a person to whom money is owing.

The amounts a debtor owes to his creditors are known as his *Liabilities*.

A *Cash Account* is a simple account with which we should all be familiar, as we all pay out and receive cash. It is prudent to keep a record of our cash receipts and payments.

A general rule for posting, *i.e.* making entries in, Ledger accounts is:

> *Debit* what comes in.
> *Credit* what goes out.

Consider the following:

Ex. 5 Enter the following transactions in your Cash Account. Balance the account and bring down the balance.

June 1. Cash in hand, £30·37.
" 5. Received cash from T. Smith, £4·48.
" 17. Paid rent, £24·75.
" 24. Paid cash to W. Brown, £7·97.
" 28. Cash sales, £67·91.
" 30. Paid into Bank, £60.

Solution:

Cash a/c.

Dr. Cr.

June 1	To Balance		£ 30·37	June 17	By Rent		£ 24·75
" 5	" T. Smith		4·48	" 24	" W. Brown		7·97
" 28	" Sales		67·91	" 30	" Bank		60·00
				" 30	" Balance	c/d	10·04
			£102·76				£102·76
July 1	To Balance	b/d	£10·04				

Notes:

1. Name of the account: Cash a/c. (a/c = Account.)
2. Left side: Dr. = Debit. Right side: Cr. = Credit.
3. Columns:
   Left: (a) Date.
      (b) Name of person or account from whom received, with prefix 'to'.
      (c) Amount.
   Right: (a) Date.
      (b) Name of person or account to whom paid, with prefix 'by'.
      (c) Amount.

4. c/d = carried down. b/d = brought down.

5. Explanation of entries $<$ Dr. Receipts.
                          Cr. Payments.

June 1. This is an amount remaining from a previous
        period.
  ,, 5. This comes in, from T. Smith.
  ,, 28.   ,,   ,,   being proceeds of sales.
  ,, 17. This goes out, to Rent a/c.
  ,, 24.   ,,   ,,   to W. Brown's a/c.
  ,, 30.   ,,   ,,   to Bank a/c.
  ,, 30. This is the difference, or amount necessary to
        balance the two columns. It is obtained by
        the *Complementary Addition* method of sub-
        traction. It is carried down to the debit
        side of the account ready for the next period,
        and is called a *Debit* balance.

We will now consider an example of a *Personal Account,*
i.e. an account in the name of a person or firm.

Ex. 6   Enter the following transactions in the account
of F. Amos. Balance the account and bring down the
balance.

Aug. 1. Balance (Cr.), £45·87.
  ,, 9. Bought goods from F. Amos, £97·58.
  ,, 11. Paid to F. Amos, Cash, £45·87.
  ,, 12. Returned goods to F. Amos, £17·51.
  ,, 30. Bought goods from F. Amos, £37.

Solution:

F. Amos.

Dr.   Cr.

Aug. 11	To Cash		£ 45·87	Aug. 1	By Balance		£ 45·87
,, 12	,, Returns		17·51	,, 9	,, Goods		97·58
,, 31	,, Balance	c/d	117·07	,, 30	,, Goods		37·00
			£180·45				£180·45
				Sept. 1	By Balance	b/d	£117·07

F. Amos has now a credit balance of £117·07. He is, therefore, our creditor for this amount.

5. Multiplication

The symbol of multiplication is a cross, ×.

The *Multiplicand* is the number to be multiplied.

The *Multiplier* is the number by which the multiplicand is multiplied.

The *Product* is the result of multiplication.

The following is an example of the usual method of the multiplication of numbers greater than 12.

Ex. 7   472 × 58.

$$
\begin{array}{rl}
472 & \text{Multiplicand} \\
58 & \text{Multiplier} \\
\hline
3\,776 & \\
23\,600 & \\
\hline
27\,376 & \text{Product}
\end{array}
$$

Mental steps:

   (1) Eight 2's = 16. Write down 6. Carry 1.
   (2) Eight 7's = 56.   +1 = 57.   Write down 7. Carry 5.
   (3) Eight 4's = 32.   +5 = 37.   Write down 37.
   (4) Place 0 in the units column, as we are multiplying by 50.
   (5) Five 2's = 10. Write down 0. Carry 1.
   (6) Five 7's = 35.   +1 = 36.   Write down 6. Carry 3.
   (7) Five 4's = 20.   +3 = 23.   Write down 23.
   (8) Add the two lines thus obtained.

6. Division

Division is the separating of a quantity into groups.

The *Dividend* is the whole amount.

The *Divisor* is the number in each group.

The *Quotient* is the number of groups.

Ex. 8 $2457 \div 9$.

This means—how many groups of 9 are there in 2457?

$$9)\overline{2457}$$
$$\mathbf{273}$$

Mental steps:

(1) Two 9's = 18. Write down 2. Carry 6.
(2) Seven 9's = 63. Write down 7. Carry 2.
(3) Three 9's = 27. Write down 3.

Exercise 1

1. Tots.

Add horizontally into the blank spaces on the right.
Then add up the four vertical columns.

A.

£	£	£	£
98·36	317·47	1 618·35	——
154·84	1 234·26	97·13	——
1 345·58	63·72	136·43	——
19·16	971·14	2 134·47	——
45·43	1 542·06	635·71	——
263·86	9·88	1 764·19	——
2 915·74	768·83	19·43	——
138·37	45·16	127·79	——
£	£	£	£

B.

£	£	£	£
717·48	12·56	1 819·94	——
1 524·56	217·47	923·58	——
16·27	1 856·36	128·47	——
856·63	621·54	1 767·68	——
1 345·53	1 211·79	11·14	——
2 347·99	345·33	2 534·08	——
17·17	1 826·97	9·99	——
256·41	17·44	813·23	——
7·57	236·84	1 232·89	——
1 927·82	58·48	426·26	——
18·89	1 234·68	327·44	——
236·63	8·89	1 349·83	——
£	£	£	£

2. Total and balance the following:

A.

£	£
3 745·48	634·61
862·78	1 349·39
1 236·96	221·77
1 837·14	18·48
5·19	615·44
27·71	2 123·24
176·46	2 754·32
135·84	Balance

£ _____   £ _____

B.

£	£
325·48	32·09
1 764·59	216·70
32·79	1 123·27
157·26	638·58
1 261·59	297·99
926·82	6·11
58·38	1 827·79
1 627·52	276·19
168·35	1 124·48
Balance	927·71

£ _____   £ _____

3. Enter the following transactions in your Cash Account. Balance the account and bring down the balance.

A. Jan. 1. Cash in hand, £11·15.
,, 4. Received cash from A. Wills, £27·50.
,, 7. Paid rent, £32·50.
,, 11. Bought goods for cash, £8·89.
,, 12. Drew cash from Bank, £45.
,, 14. Paid wages, £50.
,, 29. Cash sales, £87·79.
,, 31. Paid F. Knott, cash, £10·50.
,, 31. Paid into Bank, £60.

B. Feb. 1. Cash in hand, £34·55.
,, 6. Received cash from C. Green, £41·34.
,, 9. Paid D. Brown, cash, £3·72.
,, 14. Paid electricity bill, £28·23.
,, 14. Paid wages, £48·75.

Feb. 27. Cash sales, £100·58.
,, 28. Paid for postage, cash, £2·75.
,, 28. Paid into Bank, £90.
,, 28. Received cash from N. White, £4·50.

4. Enter the following transactions in T. Black's Account. Balance the account and bring down the balance.

March 1. Balance (Cr.), £67·50.
,, 12. Bought goods from T. Black, £68·85.
,, 14. Paid to T. Black, cash, £67·50.
,, 15. Returned goods to T. Black, £18·75.
,, 27. Bought goods from T. Black, £37·50.

5. Enter the following transactions in L. White's Account. Balance the account and bring down the balance.

June 1. Balance (Dr.), £57·72.
,, 2. Sold goods to L. White, £38·25.
,, 4. L. White returned goods, £9·50.
,, 28. L. White paid us in cash, £52·72, for the amount owing on the 1st June. We allowed him £5 discount.

6. Give the product of: A. 498 × 57.
B. 3659 × 19.
C. 13 765 × 298.

7. A. 11 408 ÷ 23.
B. 273 144 ÷ 456.
C. 86 229 ÷ 871.

8. A man shovels 160 kg of sand in 8 minutes. How much does he shovel in 3 hours?

9. 9 boys share equally 603 marbles. How many do they each receive?

10. The wages shared equally by 12 men is £1180·80. What is the wage of each man?

11. Find the value of 547 tonnes of coal at £107·50 per tonne.

12. £54·75 ÷ 75p.

13. A merchant in France bought goods valued at £458. How much will he pay in francs if £1 is worth 10·60 francs?

14. 235 men receive an average wage per week of £92. What will be the total wages bill for 26 weeks?

15. A merchant sent a consignment of goods to India to the value of £57 485. Give the value of the consignment in rupees, the exchange value of a rupee being 7½p.

16. An American firm imported 2000 cars from England at a price of £3210 for each car. Give the total value of the cars in dollars when the exchange value of the dollar is 56p. (Answer correct to the nearest dollar.)

CHAPTER 2

Fractions

1. Definitions

A *digit* is any number from 1 to 9.

A *cipher* is used to mark the absence of a digit. In a number its presence increases that portion to the left of it tenfold.

An *integral number* is one that contains an exact number of units.

An *abstract number* is one that indicates no denomination, *e.g.* 5, 12.

A *concrete number* is one that indicates a particular denomination, *e.g.* 5 kg., 12 men.

A *prime number* is one which cannot be divided, except by itself and unity, *e.g.* 5, 7, 11, 13, etc.

A *factor* or *measure* is a number which is contained an exact number of times in another number. Thus, 7 is a factor of 35.

A *composite number* is one that is composed of two or more factors, *e.g.* $35 = 5 \times 7$.

A *common factor* is a number which divides two or more numbers exactly, *e.g.* 9 is a common factor of 36 and 54.

The *highest common factor* (H.C.F.) is the greatest number which divides two or more numbers exactly, *e.g.* 18 is the highest common factor of 36 and 54.

A *multiple* is a number which contains another number exactly, *e.g.* 56 is a multiple of 7.

A *common multiple* is a number which contains two or more numbers exactly, *e.g.* 72 is a common multiple of 4, 9, and 12.

The *least common multiple* (L.C.M.) is the smallest number

which contains two or more numbers exactly, *e.g.* 36 is the L.C.M. of 4, 9, and 12.

A *fraction* is a part of a whole number. It consists of two parts.

The *denominator* is the part placed below the line. It denotes the number of parts into which the whole is divided.

The *numerator* is the part placed above the line. It denotes the number of parts to be taken. Thus $\frac{3}{8}$ means that the unit is divided into 8 parts, and 3 of them are taken.

A *proper fraction* is one whose numerator is smaller than its denominator, *e.g.* $\frac{7}{12}$.

An *improper fraction* is one whose numerator is larger than its denominator, *e.g.* $\frac{12}{7}$.

A *mixed number* consists of a whole number and a fraction, *e.g.* $1\frac{5}{7}$.

2. Addition and Subtraction

The numerator and denominator of a fraction may be multiplied or divided by the same number without altering its value.

Thus $$\frac{20}{25} = \frac{4}{5}$$

In order to add or subtract fractions it is necessary to reduce them to a common denominator.

Ex. 1
$$\frac{4}{5} + \frac{3}{4} = \frac{16+15}{20}$$
$$= \frac{31}{20}$$
$$= 1\frac{11}{20} \qquad \textbf{Ans.}$$

Ex. 2
$$\frac{7}{9} - \frac{11}{15} = \frac{35-33}{45}$$
$$= \frac{2}{45} \qquad \textbf{Ans.}$$

When several fractions are preceded by *plus* and *minus* signs, it is better to deal with the whole numbers first and then subtract *the sum* of the fractions preceded by the *minus* sign from *the sum* of the fractions preceded by the *plus* sign.

Ex. 3 $\quad 3\frac{1}{3} + 5\frac{1}{4} - 2\frac{7}{8} + 1\frac{5}{9} - \frac{2}{3} = 7 + \frac{1}{3} + \frac{1}{4} + \frac{5}{9} - \frac{7}{8} - \frac{2}{3}$

$$= 7\frac{(24+18+40)-(63+48)}{72}$$

$$= 6\frac{154-111}{72}$$

$$= 6\frac{43}{72} \qquad \textbf{Ans.}$$

3. Mental Work

Two fractions having unity for numerator may be added or subtracted at sight.

(*a*) To *add*. Multiply the denominators for a new denominator, and add the denominators for a new numerator.

Thus
$$\frac{1}{5} + \frac{1}{7} = \frac{12}{35}$$
$$\frac{1}{9} + \frac{1}{11} = \frac{20}{99}$$

(*b*) To *subtract*. Multiply the denominators for a new denominator, and subtract the smaller from the greater for a new numerator.

Thus
$$\frac{1}{5} - \frac{1}{7} = \frac{2}{35}$$
$$\frac{1}{9} - \frac{1}{11} = \frac{2}{99}$$

(*c*) These rules may be extended to fractions having any number as numerators by cross multiplication.

Thus
$$\frac{2}{3} + \frac{5}{8} = \frac{2}{3} \diagdown\!\!\!\diagup \frac{5}{8} = \frac{31}{24} = 1\frac{7}{24}$$
$$\frac{2}{3} - \frac{5}{8} = \frac{2}{3} \diagup\!\!\!\diagdown \frac{5}{8} = \frac{1}{24}$$

4. Multiplication

To multiply fractions, multiply the numerators for a new numerator, and the denominators for a new denominator.

The word 'of' is almost equivalent to the multiplication sign; it binds the quantities together as though they were in brackets (see para. 6).

Thus
$$\frac{5}{6} \text{ of } \frac{3}{4} = \frac{5}{6} \times \frac{3}{4}$$
$$= \frac{15}{24}, \text{ in lowest terms } \frac{5}{8}$$

Divide by any factors common to numerators and denominators before multiplying.

Thus $\qquad \dfrac{5}{6}$ of $\dfrac{3}{4} = \dfrac{5}{\underset{2}{6}} \times \dfrac{3}{4} = \dfrac{\mathbf{5}}{\mathbf{8}}$

The H.C.F. of two numbers may be found by a process of division. The smaller number is divided into the larger. The remainder is then divided into the previous divisor and this process is repeated until an exact divisor is found. This exact divisor is the H.C.F.

Ex. 4   Find the H.C.F. of 438 and 4453.

$$
\begin{array}{r}
438\overline{)4453}(10 \\
4380 \\
\hline
73\overline{)438}(6 \\
438 \\
\hline
\cdots
\end{array}
$$

$$\text{H.C.F.} = \mathbf{73} \qquad\qquad \textbf{Ans.}$$

Ex. 5   Reduce to lowest terms $\dfrac{273}{429}$.

(H.C.F. = 39)

$$\dfrac{273}{429} = \dfrac{7}{\mathbf{11}} \qquad \textbf{Ans.}$$

$$
\begin{array}{r}
273\overline{)429}(1 \\
273 \\
\hline
156\overline{)273}(1 \\
156 \\
\hline
117\overline{)156}(1 \\
117 \\
\hline
39\overline{)117}(3 \\
117 \\
\hline
\cdots
\end{array}
$$

A whole number may be treated as a fraction by taking *unity* as denominator.

Thus $\qquad \dfrac{4}{21}$ of $7 = \dfrac{4}{\underset{3}{21}} \times \dfrac{7}{1}$

$$= \dfrac{4}{3} = \mathbf{1\tfrac{1}{3}}$$

Ex. 6   What is the value of $\frac{5}{8}$ of £5·44?

$$\text{Value} = \pounds \frac{\overset{0·68}{\cancel{5·44}}}{1} \times \frac{5}{\cancel{8}} = \pounds\mathbf{3·40} \qquad \textbf{Ans.}$$

5. Division
To divide by a fraction, invert the divisor and multiply.

Ex. 7 $$\frac{3}{7} \div \frac{6}{11} = \frac{3}{7} \times \frac{11}{\underset{2}{\cancel{6}}}$$

$$= \frac{11}{14}. \qquad \textbf{Ans.}$$

Ex. 8 $$\frac{9}{11} \div 3 = \frac{9}{11} \div \frac{3}{1}$$

$$= \frac{\overset{3}{\cancel{9}}}{11} \times \frac{1}{3} = \frac{3}{11} \qquad \textbf{Ans.}$$

In multiplication and division, mixed numbers must be reduced to improper fractions.

Ex. 9 $$3\tfrac{3}{5} \div 2\tfrac{1}{4} = \frac{18}{5} \div \frac{9}{4}$$

$$= \frac{\overset{2}{\cancel{18}}}{5} \times \frac{4}{9} = 1\tfrac{3}{5} \qquad \textbf{Ans.}$$

6. The Use of Brackets
Brackets are used to *bind together* two or more quantities.

Thus	$16 - (7 + 8) = 1$
while	$16 - 7 + 8 = 17.$
Similarly,	$24 \div (6 + 2) = 3$
while	$24 \div 6 + 2 = 6.$

Quantities connected by the signs '×' and '÷' must be reduced to simple form before quantities connected by the signs '+' and '−'.

Ex. 10 $4\frac{1}{2} \div 2\frac{1}{4}$ of $2\frac{2}{3} = \frac{9}{2} \div \left(\frac{9}{4} \times \frac{8}{3}\right)$

$\qquad\qquad\qquad = \frac{9}{2} \div \frac{6}{1}$

$\qquad\qquad\qquad = \frac{9}{2} \times \frac{1}{6} = \frac{3}{4}$ **Ans.**

Ex. 11 $4\frac{1}{2} \div 2\frac{1}{4} \times 2\frac{2}{3} = \frac{9}{2} \times \frac{\overset{2}{\cancel{4}}}{\cancel{9}} \times \frac{8}{3}$

$\qquad\qquad\qquad\qquad = 5\frac{1}{3}$ **Ans.**

Ex. 12 $\frac{1}{2} \div \frac{2}{3} + \frac{7}{8} \times \frac{5}{7} - \frac{2}{3} = \left(\frac{1}{2} \times \frac{3}{2}\right) + \left(\frac{7}{8} \times \frac{5}{7}\right) - \frac{2}{3}$

$\qquad\qquad\qquad\qquad = \frac{3}{4} + \frac{5}{8} - \frac{2}{3}$

$\qquad\qquad\qquad\qquad = \frac{18 + 15 - 16}{24}$

$\qquad\qquad\qquad\qquad = \frac{17}{24}$ **Ans.**

7. Practical Problems

Ex. 13   3 boys share £5·40. For every 2p A receives, B receives 3p and C 4p. How much do they each receive?

Out of every 9p shared, A receives 2p or $\frac{2}{9}$.

$\qquad\qquad\qquad$ B $\quad$,, $\quad$ 3p ,, $\frac{3}{9}$.

$\qquad\qquad\qquad$ C $\quad$,, $\quad$ 4p ,, $\frac{4}{9}$.

$\left. \begin{array}{l} \text{A's share} = \frac{2}{9} \text{ of £5·40} = \textbf{£1·20} \\ \text{B's} \quad ,, \quad = \frac{3}{9} \quad ,, \quad = \textbf{£1·80} \\ \text{C's} \quad ,, \quad = \frac{4}{9} \quad ,, \quad = \textbf{£2·40} \end{array} \right\}$ **Ans.**

Ex. 14   A merchant buys a case containing 192 kg of tea. He sells $\frac{1}{3}$ on Monday and $\frac{3}{8}$ on Tuesday; on Wednesday he sells $\frac{3}{8}$ of the remainder. How many kg has he left?

On Mon. and Tues. he sells $\frac{1}{3} + \frac{3}{8} = \frac{17}{24}$

$\qquad\qquad\qquad$ Remainder $= \frac{7}{24}$.

On Wed. he sells $\frac{3}{8}$ of $\frac{7}{24}$ $\qquad = \frac{7}{64}$

$\qquad\qquad$ Amount left $= \frac{7}{24} - \frac{7}{64}$

$\qquad\qquad\qquad\qquad = \frac{56 - 21}{192}$

$\qquad\qquad\qquad\qquad = \frac{35}{192}$ of 192 kg

$\qquad\qquad\qquad\qquad = \textbf{35 kg}$ **Ans.**

Ex. 15 A machine costs £2400. It is decided to depreciate it (*i.e.* write down its value) by $\frac{1}{4}$ of its value each year. What will be its value at the end of 3 years?

$$
\begin{array}{r}
£ \\
2400 \\
\tfrac{1}{4} = \ \ 600 \\
\hline
1800 \ \ = \text{1st year} \\
\tfrac{1}{4} = \ \ 450 \\
\hline
1350 \ \ = \text{2nd year} \\
\tfrac{1}{4} = \ \ 337\cdot5 \\
\hline
1012\cdot5 = \text{3rd year}
\end{array}
$$

$\therefore$ Value = **£1012·50**   **Ans.**

Ex. 16 If a bath could be filled in 15 minutes by the cold water pipe and in 25 minutes by the hot water pipe, how long would it take to fill it if both taps were turned on together?

In 1 min   Cold water pipe fills $\frac{1}{15}$.
Hot   ,,   ,, $\frac{1}{25}$.
Together, $\frac{1}{15} + \frac{1}{25} = \frac{8}{75}$.

$\therefore$ Time taken $= \frac{75}{8} = $ **$9\frac{3}{8}$ minutes**   **Ans.**

Ex. 17 If the waste pipe could empty the above bath in $12\frac{1}{2}$ minutes, how long would it take to fill the bath if both taps were turned on and the waste plug was drawn?

In 1 min bath fills $\frac{8}{75}$ and empties $\frac{2}{25}$
$= \frac{8}{75} - \frac{2}{25} = \frac{2}{75}$.

$\therefore$ Time taken $= \frac{75}{2} = $ **$37\frac{1}{2}$ minutes**   **Ans.**

Ex. 18 One train has travelled $\frac{1}{7}$ of the distance from town A to town B, while another has travelled $\frac{1}{9}$ of the distance from town B to town A. They are then 94 km apart. What is the distance from A to B?

Distance travelled $\quad = \frac{1}{9} + \frac{1}{7}$ of whole distance.

$\qquad\qquad\qquad\quad = \frac{16}{63} \qquad\qquad$,, $\qquad\qquad$,,

$\therefore$ Remaining distance $= \frac{47}{63} = 94$ km

$\frac{1}{63}$ of whole distance $\quad = \frac{94}{47}$ km

and whole distance $\quad = \frac{94 \times 63}{47}$ km

$\qquad\qquad\qquad\qquad\quad = \mathbf{126\ km} \qquad\qquad\qquad\qquad$ **Ans.**

Exercise 2

Find the value of:

1. $\frac{1}{4} + \frac{1}{5}$.
2. $\frac{1}{4} - \frac{1}{5}$.
3. $\frac{1}{7} + \frac{1}{8}$.
4. $\frac{1}{9} - \frac{1}{11}$.
5. $\frac{2}{3} - \frac{5}{8}$.
6. $\frac{4}{15} + \frac{1}{3}$.
7. $\frac{3}{5} - \frac{2}{7}$.
8. $\frac{6}{13} - \frac{2}{5}$.
9. $\frac{1}{6} \div \frac{1}{5}$.
10. $\frac{7}{8} \times \frac{4}{5}$.
11. $1\frac{1}{2} \div \frac{6}{7}$.
12. $2\frac{1}{2} \div \frac{5}{8}$.
13. $\frac{2}{3} \times 1\frac{4}{5}$.
14. $\frac{6}{7} \div 1\frac{1}{14}$.
15. $\frac{3}{9} \times \frac{9}{14}$.
16. $3\frac{1}{2} \div \frac{7}{16}$.
17. $\frac{1}{4} + \frac{1}{2} - \frac{1}{12}$.
18. $\frac{3}{4} - \frac{5}{6} + \frac{2}{3}$.
19. $1\frac{1}{2} + 2\frac{2}{3} - 1\frac{3}{4}$.
20. $3\frac{5}{8} - 2\frac{1}{2} + \frac{5}{6}$.
21. $\frac{7}{18} \times \frac{3}{21} \div \frac{5}{9}$.
22. $7\frac{2}{5} \div 4\frac{5}{8}$.
23. $4\frac{4}{7} \times 7\frac{7}{8} \times 1\frac{1}{9}$.
24. $3\frac{1}{13} \times 2\frac{3}{5} \div (3\frac{3}{10} \times 3\frac{7}{11})$.

25. Express (a) 35 min as a fraction of 2 h.
    (b) 75 cm $\quad$,, $\quad$,, $\quad$ 9 m.
    (c) 750 g $\quad$,, $\quad$,, $\quad$ 8 kg.
    (d) 40p $\quad$,, $\quad$,, $\quad$ £3.

26. Express (a) 125 l as a fraction of 8 hl.
    (b) 625 kg $\quad$,, $\quad$,, $\quad$ 2 tonne.
    (c) 25 mm² $\quad$,, $\quad$,, $\quad$ 4 m².
    (d) 25p $\quad$,, $\quad$,, $\quad$ £8.

27. Simplify (a) $\frac{9}{13} - (\frac{1}{4} + \frac{1}{3})$.
    (b) $\frac{9}{13} - \frac{1}{4} + \frac{1}{3}$.

28. Simplify (a) $\frac{1}{12} - \frac{1}{15} + \frac{2}{3} - \frac{3}{4} + \frac{1}{10} - \frac{1}{30}$.
    (b) $4 - 2\frac{5}{13} + 18\frac{3}{7} - 19$.
    (c) $3\frac{1}{2} - 1\frac{2}{3} + \frac{5}{8} - 2\frac{5}{6} + 1\frac{3}{4}$.
    (d) $3\frac{3}{4} - \frac{2}{3} \times \frac{3}{4} + \frac{7}{8} \div 1\frac{3}{4}$.

29. Find the H.C.F. of: (*a*) 399 and 741.
    (*b*) 310 ,, 1054.
    (*c*) 686 ,, 2940.
    (*d*) 762 ,, 4719.

30. Reduce to lowest terms:
    (*a*) $\frac{148}{629}$, (*b*) $\frac{318}{583}$, (*c*) $\frac{374}{561}$, (*d*) $\frac{1253}{2864}$.

31. A road is $\frac{7}{16}$ of a km long. Give its length in metres.

32. A man's wages are $246.40 for a week of 44 hours. What is the rate of pay per hour?

33. The distance between two wickets (20 m) was measured by a metre measure which was 2 cm too short. What was the actual distance between the wickets?

34. £560 is $\frac{4}{9}$ the value of a car. What is the total value of the car?

35. £360 is divided between A, B, and C. A receives $\frac{5}{8}$, B $\frac{1}{3}$, and C the remainder. How much does C receive?

36. A grocer bought a case of tea containing 180 kg. He sold $\frac{1}{3}$ on Monday, $\frac{1}{4}$ on Tuesday, and $\frac{1}{5}$ on Wednesday. How many kg had he left?

37. It is decided to depreciate a machine which cost £500 by $\frac{1}{10}$ of its diminishing value at the end of each year. What will be its value at the end of 3 years?

38. 3 boys share £1·96. For every 2p A receives, B receives 5p and C 7p. How much do they each receive?

39. A man pays £1008 a year rent and rates. If the rates are $\frac{2}{5}$ of the rent, how much will he pay monthly for rent?

40. A earns $\frac{3}{4}$ the wages of B, and B earns $\frac{3}{4}$ the wages of C. What will A earn if C earns £120?

41. A merchant marks his goods so that $\frac{1}{8}$ of the marked price is profit. What will be: (*a*) cost price of an article marked £5·20; (*b*) the marked price of an article which cost him £3·15?

42. The coal consumption of the boiler of a 500 horse-power engine is 12 tonnes of coal in 24 hours. Give the coal consumption per horse-power hour.

Note: Fractions are sometimes printed with a sloping line '/', called a *solidus*.

Thus: $\frac{2}{7}$ may be printed 2/7.

$\frac{7}{2}$,,   ,,   7/2.

CHAPTER 3

Decimal Fractions

1. Addition and Subtraction

A *Decimal Fraction* is a fraction which has a denominator of 10 or a power of 10, *e.g.* $\frac{7}{10}$, $\frac{7}{100}$, $\frac{7}{1000}$. It is not usual to write decimal fractions with a denominator, but to write only the numerator preceded by a dot. Thus, the above would be written 0·7, 0·07, 0·007.

Our system of notation (*i.e.* counting) is a *decimal system*, and the value of a digit composing a number is one-tenth of the value it would possess were it written one place to the left.

thousands	hundreds	tens	units	tenths	hundredths
7	2	3	9	5	6

The quantity would be written 7239·56.

When the fraction is a decimal fraction, the notation is *continuous*. A *dot* is used to mark the end of the whole number and the commencement of the fractional part.

Operations are performed on the fractional part in exactly the same way as on the whole numbers.

In addition and subtraction, therefore, it is necessary to keep the figures correctly aligned in their place values, just as we do when adding or subtracting whole numbers only.

Ex. 1   235·07 + 8·3214 + 0·0036 + 137·9.

$$
\begin{array}{r}
235·07 \\
8·3214 \\
0·0036 \\
137·9 \\
\hline
\mathbf{381·295}
\end{array}
$$
                                        Ans.

Note: Ciphers to the right of the last digit of a decimal have no value.

Ex. 2   238·46 − 127·6327.

$$\begin{array}{r} 238\cdot46 \\ 127\cdot6327 \\ \hline \mathbf{110\cdot8273} \end{array}$$ **Ans.**

2. Multiplication

In multiplication the product remains the same if one of the two factors is *multiplied* and the other factor is *divided* by the same number.

Thus $53 \times 7 = 5\cdot3 \times 70 = \mathbf{371}$

$530 \times 0\cdot7 = 0\cdot53 \times 700 = \mathbf{371}$

To multiply decimals, the numbers may be treated as whole numbers and the product corrected by marking off as many decimal places as there are in the multiplier and the multiplicand.

In order to make sure that the decimal point has been placed correctly, obtain a rough estimate of the answer before working the problem.

Ex. 3   54·35 × 0·038.

(R.A. = 5 × 0·4 = 2.)   (R.A. = rough answer.)

$$\begin{array}{r} 5435 \\ 38 \\ \hline 43480 \\ 16305 \\ \hline 206530 \end{array}$$ (Mark off 5 decimal places.)

2·0653   **Ans.**

Ex. 4   0·586 × 5·7.

$$(R.A. = 0·6 \times 6 = 3·6.)$$

$$
\begin{array}{r}
586 \\
57 \\
\hline
4102 \\
29300 \\
\hline
33402 \\
\end{array}
$$
    (Mark off 4 decimal places.)

3·3402   **Ans.**

3. Division

The quotient remains the same in division if both *divisor and dividend* are multiplied or divided by the same number.

If, therefore, the divisor contains a decimal fraction:

  (1) Change the divisor to a whole number by moving the decimal point to the *right*.
  (2) Move the decimal point the same number of places to the *right* in the dividend.
  (3) Estimate the answer.
  (4) Divide in the usual way.

Ex. 5 $0·07392 \div 0·0056 = \dfrac{739·2}{56}.$ 0·0739|2
                                                          0·0056|

$$(R.A. = 72 \div 6 = 12.)$$

$$
\begin{array}{r}
13·2 \\
56)\overline{739·2(} \\
56 \\
\hline
179 \\
168 \\
\hline
112 \\
\end{array}
$$

13·2   **Ans.**

Ex. 6 $3 \cdot 948 \div 47 = \mathbf{0 \cdot 084}$ **Ans.**

$$
\begin{array}{r}
0 \cdot 084 \\
47)\overline{3 \cdot 948}(\\
3\ 76 \\
\hline
188
\end{array}
$$

Ex. 7 $10\ 983 \div 0 \cdot 523 = \mathbf{21\ 000}$ **Ans.**

$$\left(\text{R.A.} = \frac{100\ 000}{5} = 20\ 000. \right)$$

$$
\begin{array}{r}
21 \\
523)\overline{10983}(\\
1046 \\
\hline
523
\end{array}
$$

A number may be *divided* by a round number (*i.e.* 10, 100, 1000, etc.) by moving the decimal point as many places to the *left* as there are ciphers in the round number.

A number may be *multiplied* by a round number by moving the decimal point as many places to the *right* as there are ciphers in the round number.

Thus

$$5387 \div 100 = \mathbf{53 \cdot 87}$$
$$438 \cdot 2 \div 1000 = \mathbf{0 \cdot 4382}$$
$$3 \cdot 047 \times 1000 = \mathbf{3047}$$
$$5 \cdot 07 \times 1000 = \mathbf{5070}$$
$$0 \cdot 003\ 76 \times 1000 = \mathbf{3 \cdot 76}$$

Ex. 8   Simplify $\dfrac{0 \cdot 0864 \times 0 \cdot 385}{15 \cdot 84 \times 0 \cdot 001\ 05}$

$$= \frac{\overset{6}{8\cancel{6}4} \times 3\overset{11}{\cancel{8}5}}{\underset{144}{15\cancel{8}4} \times \underset{3}{\cancel{105}}}$$

$$= \mathbf{2} \qquad\qquad \textbf{Ans.}$$

Note: The numerator and denominator have been converted to whole numbers by moving the decimal point 7 places to the *right* in each case.

4. To Convert a Decimal Fraction to a Vulgar Fraction

(1) Use the figures of the decimal fraction as *numerator* of the vulgar fraction.

(2) For the denominator, write down 1 and to the right of this as many ciphers as there are ciphers and digits to the right of the decimal point.

Ex. 9
$\quad$ (a) $0.55 = \frac{55}{100} = \frac{11}{20}$

$\quad$ (b) $0.055 = \frac{55}{1000} = \frac{11}{200}$

$\quad$ (c) $0.204 = \frac{204}{1000} = \frac{51}{250}$

$\quad$ (d) $0.060 = \frac{6}{100} = \frac{3}{50}$

5. To Convert a Vulgar Fraction to a Decimal Fraction

Divide the numerator by the denominator.

Ex. 10
$\quad$ (a) $\frac{3}{5} = \mathbf{0.6}$

$\quad$ (b) $\frac{4}{9} = \mathbf{0.4}$

$\quad$ (c) $\frac{7}{16} = \mathbf{0.4375}$

Ex. 11 What is the greatest number of books, each 16 mm thick, that can be placed on a shelf 1 m long?

$$\text{Number} = \frac{1000}{16} = \frac{125}{2}$$
$$= \mathbf{62} \qquad \textbf{Ans.}$$

Ex. 12 Express 14p as the decimal of 70p.
$$\frac{14}{70} = \frac{1}{5} = \mathbf{0.2} \qquad \textbf{Ans.}$$

Ex. 13 Express 375 kg as the decimal of 2 tonnes.
$$\frac{375}{2000} = \frac{187.5}{1000} = \mathbf{0.1875} \qquad \textbf{Ans.}$$

Ex. 14   Express 75 g as the decimal of 1 kg.

$$\tfrac{75}{1000} = \mathbf{0 \cdot 075} \qquad \textbf{Ans.}$$

Ex. 15   A tank full of spirit lost 0·03 of its contents by evaporation and then 11 litres were drawn off, leaving the tank ¾ full. How many litres did the tank hold when full?

Evaporation 0·03 leaves 0·97.
¾ left = 0·75.
Amount drawn off = 0·97 − 0·75 = 0·22.
∴ 11 litres = 0·22 of contents

$$\text{and contents} = \frac{11}{0 \cdot 22} = \frac{1}{0 \cdot 02} = \frac{100}{2} = \mathbf{50 \; litres} \qquad \textbf{Ans.}$$

Ex. 16   How many metres are there in 0·566 25 of a km?

Number = 0·566 25 × 1000 = **566·25**   **Ans.**

Ex. 17   Two-fifths of a pole is in the ground, 0·125 is in the water, and 4·75 m is above water. Find the length of the pole.

$$\tfrac{2}{5} = 0 \cdot 4. \qquad 0 \cdot 4 + 0 \cdot 125 = 0 \cdot 525.$$

Decimal of pole above water = 1 − 0·525 = 0·475.
Fraction of pole above water = $\tfrac{475}{1000} = \tfrac{19}{40}$.
$\tfrac{19}{40}$ of pole = 4·75 m = $\tfrac{19}{4}$ m.
∴ Length = $(\tfrac{19}{4} \times \tfrac{40}{19})$ m = **10 m**   **Ans.**

Ex. 18   75 m of rubber tubing weighs 125 kg. Find: (a) the weight of 1 m length, (b) the weight of 1 km length, (c) the length which weighs 1 tonne.

Weight of 1 m   = $\tfrac{125}{75}$ kg = $1\tfrac{2}{3}$ kġ   **Ans.** (a)

Weight of 1 km = $(1000 \times 1\tfrac{2}{3})$ kg

$$= \frac{1000 \times 5}{3} \text{ kg}$$

$$= 1667 \text{ kg}$$

$$= \mathbf{1 \cdot 667 \; t} \qquad \textbf{Ans.} (b)$$

$$\text{Length of 1 t weight} = \frac{1000}{1\frac{2}{3}} \text{ m} = \frac{1000 \times 6}{10}\text{m}$$
$$= \textbf{600 m} \qquad \textbf{Ans. } (c)$$

Ex. 19   How many pieces of wood, 55 mm long, can be cut from a plank, 1 m long, if the saw cut is 1·25 mm, and what will be the length of the piece left over?

Length required for each piece = (5·5 + 0·125) cm = 5·625 cm.

$$\text{Number of pieces} = \frac{100}{5\cdot625}$$
$$= \textbf{17}$$
$$\left.\text{Piece left over} \quad = \textbf{4·375 cm}\right\}\textbf{Ans.}$$

$$\begin{array}{r} 17 \\ 5\cdot625)\overline{100\cdot00(} \\ \underline{56\cdot25} \\ 43\cdot750 \\ \underline{39\cdot375} \\ 4\cdot375 \end{array}$$

Ex. 20   A retailer makes a profit of 0·225 on selling price. What will be the selling price of an article which costs him £3·87½?

(Let C.P. = cost price, S.P. = selling price.)
If S.P. is 1 unit, C.P. = 1 − 0·225 = 0·775.

$$\therefore \text{S.P.} = £\frac{31}{8} \times \frac{1}{0\cdot775} = £\frac{31}{8} \times \frac{1000}{775} = \textbf{£5} \qquad \textbf{Ans.}$$

Ex. 21   A retailer makes a profit of 0·375 on cost price. What will be the cost price of an article he sells for £8·25?

If C.P. is 1 unit, S.P. = 1 + 0·375 = 1·375.

$$\therefore \qquad \text{C.P.} = £\frac{33}{4} \times \frac{1}{1\cdot375}.$$

$$= £\frac{33}{4} \times \frac{1000}{1375} = \textbf{£6} \qquad \textbf{Ans.}$$

Ex. 22   A man buys an article for £2·50 and sells it for £4. Express his profit as a decimal of: (*a*) cost price, (*b*) selling price.

$$\text{Profit} \qquad = £4 - £2·50 = £1·50.$$

$$\text{,, of C.P.} = \frac{1\frac{1}{2}}{2\frac{1}{2}} = \frac{3}{5} = \mathbf{0·6} \qquad \textbf{Ans.} \ (a)$$

$$\text{,, ,, S.P.} = \frac{1\frac{1}{2}}{4} = \frac{3}{8} = \mathbf{0·375} \qquad \textbf{Ans.} \ (b)$$

Exercise 3

1. 0·35 × 8.   2. 6 × 0·55.   3. 5·4 × 9.
4. 0·075 × 40.   5. 0·015 × 16.   6. 1·8 × 2·5.
7. 5·6 ÷ 7.   8. 8 ÷ 0·4.   9. 0·8 ÷ 4.
10. 6·4 ÷ 0·08.   11. 0·06 ÷ 12.   12. 63 ÷ 0·07.
13. 56·4 + 1872·08 + 0·092 + 56·12.
14. 0·27 + 186 + 3·456 + 192·5.
15. 183·8 + 0·0912 + 18·7 + 1·0032.
16. 187·5 − 96·038.
17. 16·027 − 11·5.
18. 0·038 − 0·0038.

19. Simplify: (*a*) $\dfrac{0·0072 \times 0·15}{1·8 \times 0·000\,45}$.

    (*b*) $\dfrac{0·001\,44 \times 15}{0·06 \times 0·072}$.

    (*c*) $\dfrac{1·82 \times 2·5}{0·35 \times 13}$.

20. How many metres are there in 0·316 25 of a km?

21. A road is 0·075 of a km long. What is its distance in metres?

22. Express a speed of 18 km an hour in metres per second.

23. A tank full of spirit lost 0·05 of its contents by evaporation. 34 litres were then drawn off, leaving the tank

two-thirds full. How many litres did the tank hold when full?

24. Two-sevenths of a pole is in the ground, 0·375 is in water, and 9·5 m is above water. Find the length of the pole.

25. How many pieces of wood, 8·45 cm long, can be cut from a plank, 1·4 m long, if the saw cut is 0·15 cm?

26. 100 m of rubber tubing weighs 200 kg. Find: (*a*) the weight of 1 m length, (*b*) the weight of 1 km length, (*c*) the length of 1 tonne in weight.

27. A shopkeeper makes a profit of 0·275 on selling price. What will be the selling price of an article which costs him £2·90?

28. A shopkeeper makes a profit of 0·225 on cost price. What will be the cost price of an article he sells for £2·45?

29. What will 10 kg of cheese cost if 800 g cost £2·28?

30. A man buys an article for £2 and sells it for £2·50. Express his profit as the decimal of: (*a*) cost price, (*b*) selling price.

CHAPTER 4

The Metric System

SI is the adopted abbreviation for Système International d'Unités (International System of Units), the modern form of the metric system adopted by the Eleventh General Conference of Weights and Measures (C.G.P.M.) in 1960, and endorsed by the International Organisation for Standardisation. The metric system is used nearly everywhere for precise measurements in science and is now being adopted throughout most of the world. It is the recognised international language of measurement.

The metric system is a decimal system. The multiples and submultiples of any unit are in powers of ten and are formed by means of prefixes which have standard names.

The following are the names of the prefixes:

Prefix	Symbol	Factor by which the unit is multiplied		
mega	M	10^6 =		1 000 000
kilo	k	10^3 =		1000
hecto	h	10^2 =		100
deca	da	10^1 =		10
deci	d	10^{-1} =	$\frac{1}{10}$ =	0·1
centi	c	10^{-2} =	$\frac{1}{100}$ =	0·01
milli	m	10^{-3} =	$\frac{1}{1000}$ =	0·001
micro	μ	10^{-6} =	$\frac{1}{1000000}$ =	0·000 001

In many countries the comma is used as a decimal marker for metric dimensions. Under the SI system, therefore, the digits of large numbers are separated into groups of three by a small gap, *not* by a comma as in the British system. A sequence of four figures should *not* be broken into groups, e.g. 4568·0793. The point will, however, continue to be the decimal marker in the United Kingdom for all purposes.

A prefix applied to a unit is subject to any applied power, *e.g.*:

$$1 \text{ mm}^3 = 1 \text{ (mm)}^3 \qquad = 10^{-9} \text{ m}^3 = \frac{\text{m}^3}{10^9}$$

not $\quad 1 \text{ m (m)}^3 \qquad = 10^{-3} \text{ m}^3 = \frac{\text{m}^3}{10^3}.$

Only one multiplying prefix is applied at one time to a given unit. Thus: one thousand kilogrammes is *not* written 1 kilokilogramme but 1 megagramme (Mg). Similarly, one thousandth of a millimetre is *not* written 1 millimillimetre but 1 micrometre (μ).

The SI unit of length is the *metre*.

The following is a table of length:

10 millimetres (mm)	=	1 centimetre
10 centimetres (cm)	=	1 decimetre
10 decimetres (dm)	=	1 metre
10 metres (m)	=	1 decametre
10 decametres (dam)	=	1 hectometre
10 hectometres (hm)	=	1 kilometre (km)

The recommended multiples and submultiples for general use are: km, m, cm, mm.

Approximate equivalents

1 km	= 0·6214 mile	1 mile	= 1·61 km
1 m	= 1·094 yd	1 yd	= 0·9144 m
1 mm	= 0·039 37 in	1 ft	= 0·3048 m
11 m	= 12 yd	1 in	= 2·54 cm
8 km	= 5 miles		

The tables of *area* are made by squaring, and those of *volume* by cubing, the units of length.

Approximate equivalents

Area:	1 km²	= 0·386 mile²	1 mile²	= 2·59 km²
	1 m²	= 1·196 yd²	1 acre	= 0·4047 ha
	1 cm²	= 0·155 in²	1 yd²	= 0·836 m²
Land:	100 m²	= 1 are (a)	1 ft²	= 929 cm²
	100 a	= 1 hectare (ha)	1 in²	= 6·45 cm²
	100 ha	= 1 km²		
	1 ha	= 2·47 acres		
Volume:	1 m³	= 1·308 yd³	1 yd³	= 0·765 m³
	1 dm³	= 0·0353 ft³	1 ft³	= 28·3 dm³
	1 cm³	= 0·061 in³	1 in³	= 16·4 cm³
Capacity:	1 dm³	= 1 litre (l)		

Approximate equivalents

1 l	= 1·76 pt	1 pt	= 0·568 l
	= 0·22 gal	1 gal	= 4·546 l

Mass

The SI unit of mass is the *kilogramme*.

1000 kg = 1 Mg = 1 t (tonne or metric ton).

Approximate equivalents

1 t	= 2205 lb	1 ton	= 1·016 t
	= 0·984 ton	1 cwt	= 50·8 kg
1 kg	= 2·205 lb	1 lb	= 0·454 kg
1 g	= 0·0353 oz	1 oz	= 28·35 g
	= 15·43 gr (grain)		

Mass and Weight

A kilogramme-mass is a unit used for describing the quantity of matter in a body, and it depends only upon the total number of atoms in the body.

A kilogramme-weight is a unit of force used for describing the force with which a body is pulled downwards in the earth's gravitational field, and it depends both on the mass of the body and on the strength of the gravitational field. Given M = mass, W = weight, g = gravity, then:

$$W = M \times g.$$

The greater the distance of a body from the surface of the earth, the smaller is the gravitational acceleration and, as there is no change in its mass, it will be seen from the above equation that its weight will be less.

A mass is commonly referred to as 'a weight'. As equal masses in the same place have equal weights, the one word 'kilogramme' to describe a unit of two different quantities 'mass' and 'force' is used without ambiguity. It is essential, however, that they are easily distinguishable when expressing precise relationships in a properly defined system of units.

Ex. 1   Reduce to kilometres: (*a*) 7600 m, (*b*) 457 000 cm.

(*a*) 7600 m   = **7·6 km**   **Ans.** (*a*)

(*b*) 457 000 cm = **4·57 km**   **Ans.** (*b*)

Ex. 2   Express in square centimetres: (*a*) 5·385 m², (*b*) 6478 mm².

   (*a*) 5·385 m² = **53 850 cm²**   **Ans.** (*a*)

   (*b*) 6478 m² = **64·78 cm²**   **Ans.** (*b*)

Ex. 3   Express in grammes: (*a*) 47·5 kg, (*b*) 4870 mg.

   (*a*) 47·5 kg = **47 500 ġ**   **Ans.** (*a*)

   (*b*) 4870 mg = **4·87 ġ**   **Ans.** (*b*)

Ex. 4   Convert 45 km to miles (8 km = 5 miles).

$$45 \text{ km} = \frac{45 \times 5}{8} \text{ miles} \qquad \left[\frac{8)225}{28\frac{1}{8}} \right]$$

$$= 28\tfrac{1}{8} \textbf{ miles} \qquad\qquad \textbf{Ans.}$$

Ex. 5   Convert 45 miles to kilometres.

$$45 \text{ miles} = \frac{45 \times 8}{5} \text{ km}$$

$$= \textbf{72 km} \qquad\qquad \textbf{Ans.}$$

Ex. 6   Convert 5 cwt to kilogrammes (1 kg = 2·2 lb).

$$5 \text{ cwt} = 560 \text{ lb} \qquad \left[\frac{11)2800}{254·545} \right]$$

$$= \frac{560}{2·2} \text{ kg}$$

$$= \textbf{254·545 kġ} \qquad\qquad \textbf{Ans.}$$

Ex. 7   Convert 70 kg to pounds

$$70 \text{ kg} = 70 \times 2·2 \text{ lb}$$

$$= \textbf{154 lb} \qquad\qquad \textbf{Ans.}$$

Ex. 8   A petrol tank holds 10 gal. How many litres will it hold? (Answer to the nearest litre.)

$$10 \text{ gal} = \frac{10 \times 8}{\underset{0·22}{1·76}}$$

$$= \textbf{45 litres} \qquad\qquad \textbf{Ans.}$$

Ex. 9   A light aircraft travels at 350 miles an hour. Express its speed in metres per second (8 km = 5 miles).

$$\text{Speed} = \frac{350 \times \overset{4}{\cancel{8}} \times \cancel{1000}}{\cancel{5} \times \underset{\underset{9}{18}}{\cancel{3600}}} \text{m/s} \quad \left[\frac{9)1400}{155\cdot 5}\right]$$

$$= 155\cdot 6$$

Approx. = **156 m/s**   **Ans.**

Ex. 10   Convert a speed of 156 metres per second to miles per hour. From your answer give the error in miles per hour in the approximation in Ex. 9.

$$156 \text{ m/s} = \frac{\overset{39}{\cancel{156}} \times \cancel{3600} \times \overset{9}{\cancel{5}}}{\cancel{1000} \times 8} \text{ miles/h}$$

$$= \textbf{351 miles/h} \qquad \textbf{Ans. (i)}$$

$$\text{Error} = \textbf{1 mile/h} \qquad \textbf{Ans. (ii)}$$

Exercise 4

1. Express in kilogrammes: (*a*) 3461 grammes, (*b*) 56 427 milligrammes.

2. Express in square centimetres: (*a*) 47·5 m², (*b*) 365 mm².

3. Find the capacity in litres of a petrol tank which holds 12 gal (1 litre = 1·76 pints).

4. A boy cycles at the rate of 9 miles an hour. Find his speed in kilometres per hour.

5. Convert 125 litres to gallons.

6. How many tons are there in 56 tonnes? (1 kg = 2·204 lb).

7. The distance between two French towns is 90 km. What is the distance in miles?

8. Convert 3 cwt to kilogrammes (1 kg = 2·2 lb).

9. A car travels at the rate of 35 miles an hour. How long, to the nearest minute, will it take to travel 75 km?

10. Express 484 gal in hectolitres.

11. Convert 500 acres to hectares.

12. A cyclist rode 54 km in $2\frac{1}{4}$ hours. Find his speed in miles per hour.

13. 500 litres of vinegar is put into bottles. Each bottle holds $\frac{1}{2}$ pt and is sold at 11p. What was the total amount received?

14. Express an acre as the decimal of a hectare (1 hectare = 2·47 acres).

15. An aeroplane travels 2000 km in $2\frac{1}{4}$ hours. Express its speed in feet per second.

16. Express a speed of 30 metres per second in miles per hour.

17. A cyclist rode 100 km in $5\frac{3}{4}$ hours. Give his speed in miles per hour.

18. A motorist travels 52 miles in 2 h 10 min. Find his speed in kilometres per hour.

19. 100 litres of wine is bottled. 8 pints fill 6 bottles. The wine is sold at £2·27 a bottle. Find the total amount received.

20. By how many lb would the answer to Q. 6 be increased by taking the approximation of 1 kg = 2·204 62 lb?

21. Convert the following speed limits from miles/h to km/h (take 8 km = 5 miles):
    (*a*) 30, 40, and 70 miles/h.
    (*b*) Round off these speeds to the nearest 10 km/h.

CHAPTER 5

Approximations—Contracted Multiplication and Division

1. Approximations

An *approximate* result is one which is sufficiently correct to serve the purpose for which it is required.

In many commercial calculations, for example, a sum of money is expressed with sufficient accuracy if it is correct to the nearest new penny.

The distance between towns remotely situated is sufficiently accurate if given to the nearest kilometre, but if we were measuring out a cricket pitch we should want the distance to be correct within a centimetre or so.

In giving an approximate answer, it should be remembered that when the digit to the *right* of the last digit required is 5 or *more*, the last digit required is increased by 1.

A *significant digit* is usually considered to be any number from 1 to 9, but when a cipher occurs between 2 significant digits or takes the place of the last significant digit it is regarded as significant.

Ex. 1   Give the following correct to 3 *significant digits*:

(a) 47 135   = **47 100**

(b) 50 923   = **50 900**

(c) 28·638   = **28·6**

(d) 0·063 729 = **0·0637**

(e) 0·010 57   = **0·0106**

(f) 25·027   = **25·0**

(g) 49 047   = **49 000**

Ex. 2   Give the following correct to 4 *decimal places*:

(*a*) 5·307 246 5 = **5·3072**

(*b*) 0·005 673 4 = **0·0057**

(*c*) 6·035 850 2 = **6·0359**

Note the following:

8652·7354 to nearest	thousandth	= 8652·735.
,,	,, hundredth	= 8652·74.
,,	,, tenth	= 8652·7.
,,	,, whole number	= 8653.
,,	,, ten	= 8650.
,,	,, hundred	= 8700.
,,	,, thousand	= 9000.

2. Contracted Multiplication

Ex. 3   4·583 257 91 × 61·5243. (Correct to 1 decimal place.)

If we were to multiply by the ordinary method, our product would contain 12 places of decimals. It is evident therefore, that we need a contracted method.

Rules:

(1) Obtain a rough answer.

(2) Convert multiplier to *standard form*, *i.e.* 1 whole number.

(3) Place multiplier under multiplicand, point under point.

(4) Rule a vertical line 2 places to the *right* of the place of accuracy required. This is for greater accuracy.

(5) Commence multiplying with the *left-hand* figure of the *multiplier* and the *right-hand* figure of the *multiplicand*. Ignore any figures in the multiplicand which are to the right of the vertical line *except* the first figure, which is multiplied and the carrying figure is added to the first partial product.

The first figure of each partial product is placed to the left of the vertical line.

(R.A. = 46 × 6 = 276.)

$$
\begin{array}{r|l}
4\dot5\cdot8\dot3\dot2 & 5791 \\
6\cdot1\dot5\dot2 & 43 \\
\hline
274\cdot995 & \\
4\cdot583 & \\
2\cdot292 & \\
92 & \\
18 & \\
1 & \\
\hline
281\cdot981 & = \mathbf{282\cdot0} \qquad \textbf{Ans.}
\end{array}
$$

Note: Proceed to *right* with figures in multiplier.

   ,, *left* ,, ,, multiplicand.

To prevent error, place a dot over each figure or cross it out as it is used.

1st line.	5 × 6 = 30.	Carry 3.	
	2 × 6 = 12.	+3 = 15.	Write down 5 and continue the line.
2nd line.	2 × 1 = 2.	Nothing to carry. Write down 4583.	
3rd line.	3 × 5 = 15.	Carry 2.	
	8 × 5 = 40.	+2 = 42.	Write down 2 and continue the line.
4th line.	8 × 2 = 16.	Carry 2.	
	5 × 2 = 10.	+2 = 12.	Write down 2 and continue the line.
5th line.	5 × 4 = 20.	Carry 2.	
	4 × 4 = 16.	+2 = 18.	Write this down.
6th line.	4 × 3 = 12.	Write down 1, the carrying figure.	

Ex. 4 396 278 × 3065 (to the nearest million).

   (R.A. = 4 000 000 × 300 = 1 200 000 000.)

Answer will contain 4 significant digits, so we take 6 for greater accuracy. Draw the vertical line 5 places to the right.

$$\begin{array}{r} 3\dot{9}6\dot{2}\dot{7}|8 \\ 3\dot{0}65| \\ \hline 118883| \\ 2377| \\ 198| \\ \hline 121458 \end{array}$$ **1 215 000 000**   **Ans.**

Note: When a cipher occurs in the multiplier, place a dot over the cipher and the next figure to the left in the multiplicand, *i.e.* the 2.

3. Contracted Division

Rules:

    (1) Obtain a rough answer.

    (2) The rough answer will place the decimal point. We are then able to find the number of significant digits required.

    (3) Take one more digit in the divisor than the number required in the quotient.

    (4) Take sufficient digits in the dividend to complete the first division.

    (5) Instead of bringing down ciphers after each division, cross off the right-hand digit in the divisor, but multiply the last figure crossed off in order to obtain a carrying figure for the next product.

Ex. 5   0·006 837 24 ÷ 0·173 82 (to 4 decimal places).

$$\left(\text{R.A.} = \frac{0{\cdot}68}{17} = 0{\cdot}04.\right)$$

Number of significant digits is 3.

$$\begin{array}{r} 17\cancel{38})6837(3934 \\ 5214 \\ \hline 1623 \\ 1564 \\ \hline 59 \\ 52 \\ \hline 7 \end{array}$$

       0·0393   **Ans.**

Ex. 6   837 058·7 ÷ 39·428 52 (to the nearest hundred).

$$\left(\text{R.A.} = \frac{84\ 000}{4} = 21\ 000.\right)$$

$$3942)8370(2123$$
$$\underline{7884}$$
$$486$$
$$\underline{394}$$
$$92$$
$$\underline{79}$$
$$13$$

21 200   **Ans.**

4. Practical Problems

Ex. 7   9·1275 tonnes of coal was bought for £972·08. Find, *to the nearest penny*, the cost per tonne.

$$\left(\text{R.A.} = £\frac{972}{9} = £108.\right)$$

Significant digits = 5.

$$91275)97208(1065$$
$$\underline{91275}$$
$$5933$$
$$\underline{5476}$$
$$457$$
$$\underline{456}$$
$$1$$

Cost per tonne = **£106·50**   **Ans.**

Ex. 8   The circumference of a circle is approximately 3·141 59 times the diameter. Find, to 1 decimal place, the circumference of a circle of diameter 8·67 metres.

(R.A. = 9 × 3 = 27 m.)

$$\begin{array}{r|l} 3 \cdot \dot{1}4\dot{1} & 59 \\ 8\ 6\dot{7}\dot{0} & \\ \hline 25 \cdot 132 & \\ 1 \cdot 885 & \\ 220 & \\ \hline 27 \cdot 237 & \end{array}$$

Circumference = **27·2 m**   **Ans.**

Exercise 5

1. 15·0678 × 3·2537   correct to 2 decimal places.
2. 38·5482 × 14·063 78   ,,   2   ,,
3. 27·0862 × 35·3076   ,,   3   ,,
4. 6·897 23 × 275·237   ,,   3   ,,
5. 5·9307 × 27·286   ,,   the nearest tenth.
6. 32·3872 × 157·063   ,,   ,,   hundredth.
7. 182·54 × 27·673   ,,   ,,   whole number.
8. 3487 × 57 406   ,,   ,,   thousand.
9. 5·4368 × 27·203   ,,   3 significant digits.
10. 27 304 × 8907   ,,   3   ,,
11. 5·837 ÷ 0·0687   ,,   3   ,,
12. 687 348 ÷ 37·2513   ,,   3   ,,
13. 0·082 354 ÷ 0·510 27   ,,   2   ,,
14. 8·235 ÷ 0·413 72   ,,   the nearest whole number.
15. 15·207 ÷ 0·820 53   ,,   ,,   integer.
16. 0·053 26 ÷ 0·002 391 7 ,,   ,,   tenth.
17. 5·326 ÷ 0·239 17   ,,   ,,   hundredth.
18. 54·8623 ÷ 87·9254   ,,   ,,   thousandth.
19. 7·2516 ÷ 8·124 32   ,,   2 decimal places.
20. 0·637 95 ÷ 9·258 37   ,,   4   ,,

21. The area of a certain district was 756 849 ha, and the population was 2 758 965. Find, to the nearest whole number, the number of people per 1000 ha.

22. Taking the circumference of a circle as 3·141 59 times the diameter, find the diameter of a circular pond which is 235 m in circumference. (Answer to the nearest 0·1 m.)

23. Find, to the nearest 0·1 m, the circumference of a circular pond which has a diameter of 146 m.

24. The distance covered by the vehicles of a transport company was 237 456 km. The gross receipts were £197 801. Find, to the nearest penny, the gross receipts per km.

CHAPTER 6

Percentages—Discounts

1. Percentages

In order to assess with certainty the relative values of fractions of different denominators it is necessary to bring them to a common denominator. If, for example, we wished to compare the three fractions $\frac{2}{3}$, $\frac{3}{4}$, $\frac{5}{7}$, we could bring them to their common denominator of 84, and we should then get $\frac{56}{84}$, $\frac{63}{84}$, $\frac{60}{84}$. This denominator would apply to these three fractions. If we had a fourth fraction, $\frac{2}{5}$, we should require a different denominator for the comparison.

For business purposes it is usual to employ a common denominator of 100, and the fractions would then be $\frac{66\frac{2}{3}}{100}$, $\frac{75}{100}$, $\frac{71\frac{3}{7}}{100}$, which would be expressed as $66\frac{2}{3}$ per cent., 75 per cent., $71\frac{3}{7}$ per cent.

The term *per cent.* (from the Latin *per centum*) means 'per 100' or 'out of 100'. £4 per cent. means £4 per £100 or $\frac{4}{100}$ of any quantity of money. The numerator of the fraction is referred to as the *rate* per cent. or just the *rate*.

Per cent. is sometimes written p.c. and often written %.

Consider the following:

(a) $\frac{4}{5} = \frac{4}{5} \times \frac{100}{1}\% = \mathbf{80\%}$

(b) $0.35 = 0.35 \times \frac{100}{1}\% = \mathbf{35\%}$

(c) $12\frac{1}{2}\% = 12\frac{1}{2}$ out of 100 $= \dfrac{12\frac{1}{2}}{100} = \frac{1}{8}$

(d) $7\frac{1}{2}\% = 7\frac{1}{2}$ out of 100 $= \mathbf{0.075}$

From the above we can formulate the following rules:

(1) To convert a decimal or vulgar fraction to a per centage: *Multiply by* 100.

(2) To convert a percentage to a decimal or vulgar fraction: *Divide by* 100.

Ex. 1 Convert $\frac{5}{8}$ to a percentage.

$$\frac{5}{8} = \frac{5}{8} \times \frac{100}{1}\% = \mathbf{62\cdot5\%} \qquad \textbf{Ans.}$$

Ex. 2 Convert 0·035 to a percentage.

$$0\cdot035 = 0\cdot035 \times 100\% = \mathbf{3\tfrac{1}{2}\%} \qquad \textbf{Ans.}$$

Ex. 3 Convert 45% to (a) a fraction, (b) a decimal.

$$45\% = \frac{45}{100} \qquad = \frac{9}{20} \qquad \textbf{Ans.} \ (a)$$
$$= 45 \div 100 = \mathbf{0\cdot45} \qquad \textbf{Ans.} \ (b)$$

To find the percentage one quantity is of another:

Express the former as a fraction of the latter and multiply by 100.

Ex. 4 What percentage of £37·50 is £5?

$$£5 = \frac{5}{\underset{3}{37\cdot5}} \times \frac{\overset{8}{100}}{1} = \mathbf{13\tfrac{1}{3}\%} \qquad \textbf{Ans.}$$

Ex. 5 The population of a village in 1958 was 2376. In 1968 it was 2769. Find the percentage increase.

Increase $= 2769 - 2376 = 393.$

$$= \frac{\overset{131}{393}}{\underset{792}{2376}} \times \frac{100}{1}\%$$

$$= \mathbf{16\cdot54\%} \qquad \textbf{Ans.}$$

$$\begin{array}{r} 16\cdot54 \\ 792)13100(\\ 792 \\ \hline 5180 \\ 4752 \\ \hline 428 \\ 396 \\ \hline 32 \end{array}$$

Ex. 6 From a coal dump of 125 tonnes, 31·25 tonnes has been withdrawn. Find the percentage of the original quantity still in the dump.

Amount remaining $= 93\frac{3}{4}$ tonnes

$$= \frac{93\frac{3}{4}}{125} \times \frac{100}{1}\%$$

$$= \frac{375}{500} \times \frac{100}{1}\%$$

$$= 75\% \qquad \textbf{Ans.}$$

To find the value of a percentage of a quantity: *Reduce the percentage to a fraction and multiply by the quantity.*

Ex. 7 Find the value of 35% of £53.

$$\text{Value} = £\frac{\overset{7}{\cancel{35}}}{\underset{20}{\cancel{100}}} \times \frac{53}{1} \qquad [371]$$

$$= \textbf{£18·55} \qquad \textbf{Ans.}$$

Ex. 8 How far is $62\frac{1}{2}\%$ of 17 km?

$$\text{Distance} = \frac{\overset{5}{\cancel{62·5}}}{\underset{8}{\cancel{100}}} \times \frac{17}{1} \text{ km}$$

$$= \textbf{10}\tfrac{5}{8} \textbf{ km or 10·625 km} \qquad \textbf{Ans.}$$

2. Discounts

Discount is a deduction from the price of an article or from an amount of money owing by one person to another.

Trade discount is discount allowed by one trader to another.

Cash discount is discount allowed in consideration of prompt payment.

Gross price is the price before discount has been deducted.

Net price is the price after discount has been deducted. In some cases, a price on which no discount will be allowed.

Discount may be expressed as a definite amount, but it is more usually expressed as a percentage.

Ex. 9 A dealer allows his customers 5p in the £ discount for cash. What will be the cash price of a radio priced at £23?

(For every £1 the customer pays 95p.)

$$\text{Cash price} = £\frac{23}{1} \times \frac{19}{20}$$

$$= £21\cdot85 \quad \text{Ans.} \quad \left[\begin{array}{c} 460 \\ 23 \\ \hline 437 \end{array}\right]$$

Or: $\qquad £23 - £1\cdot15 = £21\cdot85 \qquad$ **Ans.**

Ex. 10 A customer paid a dealer £49·87½ for a radio. 5p in the £ had been deducted for cash payment. What was the marked price of the item?

$$\text{Marked price} = £\frac{49\cdot87\frac{1}{2}}{1} \times \frac{100}{95}$$

$$= £\frac{\overset{21}{\cancel{399}}}{\underset{2}{\cancel{8}}} \times \frac{\overset{5}{\cancel{20}}}{\cancel{19}}$$

$$= £52\cdot50 \qquad \text{Ans.}$$

Ex. 11 A publisher allows a discount of 16⅔% on the catalogue prices of books. How much will a bookseller pay for 100 books at £1·50 each?

$$(100 - 16\tfrac{2}{3} = 83\tfrac{1}{3}.)$$

Catalogue price = £150.

$$\text{Bookseller pays} \quad £\frac{150}{1} \times \frac{83\frac{1}{3}}{100}$$

$$= £\frac{150}{1} \times \frac{250}{300}$$

$$= £125 \qquad \text{Ans.}$$

Or: $16\tfrac{2}{3}\% = \tfrac{1}{6}$.

$$£150 - (\tfrac{1}{6} \text{ of } £150) = £150 - £25 = £125 \qquad \text{Ans.}$$

Ex. 12 A bookseller paid £90 for a parcel of 60 books on which he had been allowed a discount of 16⅔%. What should be the catalogue price of one book?

$$\text{Catalogue price} = \pounds\frac{90}{1} \times \frac{100}{83\frac{1}{3}} \times \frac{1}{60}$$

$$= \pounds\frac{\cancel{90} \times \cancel{300}}{\cancel{60} \times \cancel{250}}{5}$$

$$= \pounds1\cdot80 \qquad\qquad \text{Ans.}$$

Ex. 13   A wholesaler allows a retailer 25% trade discount and 5% cash discount. What will be the cash price of an article, the catalogue price of which is £35?

	£
Catalogue price	35·00
Less 25% trade discount	8·75
	26·25
Less 5% cash discount	1·31
Cash price	**£24·94** **Ans.**

Note: The student should *not* add the two discounts together and deduct 30%, as he would then be taking 5% of £35 as cash discount, whereas the amount owing when the cash discount is deducted is £26·25 and it is from this amount that the cash discount is deducted.

Ex. 14   A retailer is allowed 33⅓% trade discount and 5% discount for cash. What will be the catalogue price of an article for which he pays cash £17·10?

$$\text{Catalogue price} = \pounds\frac{17\cdot1}{1} \times \frac{100}{66\frac{2}{3}} \times \frac{100}{95}$$

$$= \pounds\frac{\overset{9}{\cancel{171}} \times \cancel{10} \times \cancel{300}}{\cancel{200} \times 9\cancel{5}}{5}$$

$$= \pounds27 \qquad\qquad \text{Ans.}$$

Ex. 15   A wholesaler wishes to receive £57 for a chair after allowing the retailer 33⅓% trade discount and 5%

cash discount. What will be his catalogue price for the chair?

$$
\begin{array}{lr}
& \pounds \\
\text{Let catalogue price be} & 120 \\
\text{Less } 33\frac{1}{3}\% \text{ trade discount} & \underline{40} \\
& 80 \\
\text{Less } 5\% \text{ cash discount} & \underline{4} \\
\text{Cash price} & \underline{\pounds 76}
\end{array}
$$

$$
\text{Catalogue price } = \pounds\frac{\overset{3}{\cancel{5\!7}} \times \overset{30}{\cancel{120}}}{\underset{4}{\cancel{76}}}
$$

$$
= \textbf{£90} \qquad \textbf{Ans.}
$$

Note: As $33\frac{1}{3}\% = \frac{1}{3}$, 120 is an easier starting number, as it avoids fractions.

3. Quick Methods

With a little practice it will be found quite easy to write down the solutions of simple percentage problems at sight; that is, without any written work.

To find the value of a percentage of a quantity:

   (*a*) Multiply the quantity by the rate.
   (*b*) Place a decimal point two places to the left.

Ex. 16   5% of £48 = **£2·40**

Mental steps:

   (1) $48 \times 5 = 240$.
   (2) Decimal point two places to left $= 2·40$.

Ex. 17   7% of £15 = **£1·05**

Ex. 18   19% of £12 = **£2·28**

Ex. 19   17% of £40 = **£6·80**

    Note:
$$
\begin{aligned}
5\% &= \tfrac{1}{20}. \\
2\tfrac{1}{2}\% &= \tfrac{1}{40}. \\
10\% &= \tfrac{1}{10}. \\
12\tfrac{1}{2}\% &= \tfrac{1}{8}.
\end{aligned}
$$

Ex. 20   Write down the value of:

(a)   5% of £68.   **Answers**   **£3·40**   (a)

(b)   2½% ,, £36.   **90p**   (b)

(c)   7½% ,, £54.   **£4·05**   (c)

(d)   10% ,, £63.   **£6·30**   (d)

(e)   12½% ,, £64.   **£8**   (e)

Mental steps:

(a) £68 ÷ 20 = £3·40.
(b) £36 ÷ 40 = 90p.
(c) $\frac{1}{20}$ = £2·70. (£2·70 + ½ of £2·70) = £4·05.
(d) Take one-tenth = £6·30.
(e) Divide £64 by 8 = £8.

Exercise 6

Write down the value of each of the following:

1. 2½% of £38.		2. 5% of £48.	
3. 8% ,, £32.		4. 12% ,, £35.	
5. 7½% ,, £28.		6. 10% ,, £84.	
7. 12% ,, £65.		8. 12½% ,, £96.	
9. 17% ,, £30.		10. 29% ,, £40.	
11. 5% ,, £27.		12. 11% ,, £45.	

13. What percentage is 45p of £2·25?
14.   ,,   ,,   90p ,, £10?
15.   ,,   ,,   15p ,, £2?
16.   ,,   ,,   55p ,, £2·20?
17.   ,,   ,,   75p ,, £6?
18.   ,,   ,,   54p ,, £2·70?
19.   ,,   ,,   40p ,, £5?
20.   ,,   ,,   65p ,, £6·50?

Express the following fractions as percentages:
21. $\frac{3}{4}$.   22. $\frac{3}{8}$.   23. $\frac{4}{5}$.   24. $\frac{2}{3}$.   25. $\frac{5}{7}$.

Express the following percentages as fractions:
26. 15%.   27. 12½%.   28. 45%.   29. 30%.   30. 62½%.

31. On a house bought for £21 500 a man is required to deposit 15% of the purchase price. How much remains to be paid?

32. 750 candidates sat for an examination. 585 passed and the rest failed. Give the percentage of (a) those who passed, and (b) those who failed.

33. A man spends 92% of his income and saves £540. What is his income?

34. A tradesman's weights are found to be 2% under-weight. What would be his additional gain on the sale of 56 kg of bacon at £2·65 a kg?

35. The population of a village was 720 in 1962 and 828 in 1972. Find the percentage increase.

36. A tradesman's receipts last week amounted to £2904. In the previous week they were £2548. Find, to 2 places of decimals, the increase per cent.

37. The population of a certain town is 186 745 and the number of unemployed is 9149. Find, to the nearest tenth, the percentage unemployed.

38. A man has a bank overdraft of £354. This is 60% of his monthly salary. What is his salary per annum?

39. A dress marked at £24·60 is subject to a discount of 5% for cash payment. What will be the cash price?

40. A man paid £77·70 for a radio after 7½% discount had been deducted. What was the marked price of the radio?

41. A wholesaler allows 25% trade discount and 5% discount for cash. What will be the cash price of an article marked at £12?

42. A publisher allows a trade discount of 16⅔%. What will be the cost of 48 books at £1·50 a copy?

43. A wholesaler wishes to receive £3·70 for an article after allowing the retailer 33⅓% trade discount and 7½% cash discount. At what price will he mark the article?

44. An army lost 15% of its men by disease. It then lost 10% of the remainder in battle. The number remaining was 71 145. How many men were there originally?

CHAPTER 7

Invoices

Most of us are familiar with a tradesman's bills. This is a list of the articles bought and their prices. If we pay at the time of purchase the bill is receipted. The person who receives the money writes the word 'paid' and adds his initials or signature. If we do not pay at the time of purchase the amount for the goods is credited to our account in the tradesman's ledger.

A *Cash sale* is one in which payment for the goods is received when the goods are sold.

A *Credit sale* is one in which payment for the goods is to be made at some future time.

Similar to our tradesman's bill is a document used in business. This document is called an invoice and is sent to the purchaser when goods are dispatched.

An *Invoice* is a document giving the quantity, description, and prices of goods sold.

The following is a specimen of an invoice:

INVOICE

No. 217.

45 Bright Street,
Sheffield.
4th October, 1982.

Mr. B. Trader,
  76 Kingsway,
    London W.C.2.

Bought of V. Sharpe.

	£	£
70 Table Knives @ £1·60 ea.	112·00	297·00
1 Silver-plated Tea Set @ £185	185·00	74·25
Less 25% discount . .		
		£222·75

Ex. 1   Find the amount of the following bill:

24 eggs at 7p each.
550 g bacon at £2·65 a kg.
6 kg sugar at 40p a kg.
875 g tea at £1·56 a kg.
9 tins spaghetti at 15p each.

Solution:

	£
24 eggs at 7p	= 1·68
550 g bacon at £2·65 a kg	= 1·46
6 kg sugar at 40p a kg	= 2·40
875 g tea at £1·56 a kg	= 1·37
9 tins spaghetti at 15p	= 1·35

£8·26

Exercise 7

1. Find the cost of the following:

    6 loaves bread at 42½p a loaf.
    36 eggs at 7p each.
    500 g butter at £2·02 a kg.
    12 tins sardines at 23p each.
    6 tins salmon at 95p each.
    750 g tea at £1·60 a kg.

2. Make out an invoice for the following:

    20 pr. men's shoes at £27·75 a pr.
    24 pr plimsolls at £2·35 a pr.
    40 pr. ladies' shoes at £19·20 a pr.
    60 pr. children's sandals at £2·95 a pr.

The whole invoice subject to 15% discount.

3. Find the total cost of the following:

    1·2 kg bacon at £2·65 a kg.
    1·5 kg butter at £2·02 a kg.
    25 eggs at 7p each.
    10 kg sugar at 40p a kg.

1·25 kg cheese at £2·85 a kg.
2·75 kg ham at £4·40 a kg.

4. Make out an invoice for the following:

40 kitchen chairs at £26·50 each.
3 drawing-room suites at £550 each.
2 dining-room suites at £730 each.
10 easy chairs at £158 each.

The whole invoice subject to 20% discount.

5. Make out an invoice for the following:

40 tea services at £22·50 each.
50 cruet sets at £3·75 each.
30 salad sets at £5·75 a set.
60 teapots at £3·60 each.
20 dinner services at £62·00 each.

The whole invoice subject to 25% discount.

CHAPTER 8

Profit and Loss—Pricing Goods

If an article which cost £10 is sold for £8, it is clear that £2 has been lost on the deal, and the person selling it is said to have sold the article at a loss. If the article had been sold for £12, then the person selling it would have gained £2 on the deal and is said to have sold it at a profit. If, however, the seller had paid 25p on carriage in order to deliver the article to the buyer, then his profit would be reduced by this 25p to £1·75. We have now two different amounts as the profit and, obviously, there are two kinds of profit. So when we speak of profit we should make it quite clear to which of these two kinds of profit we refer. The following are the definitions of these two kinds of profit.

Gross Profit is the amount by which the selling price exceeds the cost price of goods sold.

Net Profit is the profit remaining after all expenses have been deducted.

Consider the following:

A buys an article for £10 and sells it for £12.
B   „   „   £8   „   „   £10.

In each case, the actual profit is £2, but

B has made a profit of £2 on an outlay of £8, while
A   „   „   „   £2   „   „   £10.

Obviously B has made a better deal, and his profit is *relatively* greater than A's. His profit expressed as a ratio of either his outlay or his return is higher than A's.

To determine *relative* gains or losses, it is usual to express profits or losses as a percentage on cost price or on selling price.

The business man usually reckons his profits as a percentage on *turnover*, or *selling price*. In solving arithmetical problems, profit is expressed as a percentage on *outlay*, or *cost price*. Unless instructed to the contrary, *give percentage profit or loss on cost price*.

We will now find the relative profits of our friends A and B.

$$\left.\begin{array}{l} \text{A's profit} = \frac{2}{10} \times \frac{100}{1}\% = 20\% \\ \text{B's \quad ,,} \quad = \frac{2}{8} \times \frac{100}{1}\% = 25\% \end{array}\right\} \text{on Cost Price.}$$

$$\left.\begin{array}{l} \text{A's profit} = \frac{2}{12} \times \frac{100}{1}\% = 16\frac{2}{3}\% \\ \text{B's \quad ,,} \quad = \frac{2}{10} \times \frac{100}{1}\% = 20\% \end{array}\right\} \text{on Selling Price.}$$

Ex. 1   A man bought an article for £25 and sold it for £20. Find his loss as a percentage: (*a*) on cost price, (*b*) on selling price.

$$\text{Loss} = £25 - £20 = £5.$$
$$\text{,, \quad on C.P.} = \frac{5}{25} \times \frac{100}{1}\% = \mathbf{20\%} \quad \textbf{Ans.} \ (a)$$
$$\text{,, \quad on S.P.} = \frac{5}{20} \times \frac{100}{1}\% = \mathbf{25\%} \quad \textbf{Ans.} \ (b)$$

Note: C.P. = Cost Price.
　　　S.P. = Selling Price.

Ex. 2   A grocer bought 28 kg of tea for £35 and sold it at £1·60 a kg. Find his percentage profit: (*a*) on outlay, (*b*) on turnover.

Profit = £44·80 − £35 = £9·80.

Profit on outlay $= \dfrac{9\cdot8 \times 100}{35}\%$

$= \mathbf{28\%} \ \textbf{Ans.} \ (a)$

Profit on turnover $= \dfrac{\overset{\cdot7}{\cancel{9\cdot8}} \times 100}{\underset{3\cdot2}{\cancel{44\cdot8}}}\%$

$= \mathbf{21\cdot88\%} \quad \textbf{Ans.} \ (b)$

$$\begin{array}{r} 32)\overline{700}(21\cdot875 \\ 64 \\ \hline 60 \\ 32 \\ \hline 280 \\ 256 \\ \hline 240 \\ 224 \\ \hline 160 \\ 160 \\ \hline \end{array}$$

Ex. 3   A firm's sales last year were £640 000. The cost of the goods sold was £500 000. The selling expenses were £100 000. Find: (*a*) the Gross Profit, and (*b*) the Net Profit as a percentage on turnover.

Gross profit = £640 000 − £500 000 = £140 000.

$$= \frac{\overset{7}{\cancel{14}}}{\underset{8}{\cancel{64}}} \times \frac{\overset{25}{\cancel{100}}}{1}\%$$

$$= 21\tfrac{7}{8}\% = \mathbf{21 \cdot 875\%} \qquad \textbf{Ans.} \ (a)$$

Net profit $= \dfrac{4}{64} \times \dfrac{100}{1}\%$

$$= 6\tfrac{1}{4}\% = \mathbf{6 \cdot 25\%} \qquad \textbf{Ans.} \ (b)$$

Ex. 4   A car was bought for £1250 and sold at a profit of 16%. Find the selling price.

(For every £100 cost, seller receives £116.)

$$\text{Selling price} = \pounds \frac{\overset{50}{\cancel{1250}} \times \overset{29}{\cancel{116}}}{\underset{4}{\cancel{100}}}$$

$$= \mathbf{\pounds 1450} \qquad \textbf{Ans.}$$

Ex. 5   A car was bought for £2400 and sold at a loss of $12\tfrac{1}{2}\%$. Find the selling price.

(For every £100 cost, seller receives £87½.)

$$\text{Selling price} = \pounds \frac{2400 \times 87 \cdot 5}{100}$$

$$= \mathbf{\pounds 2100} \qquad \textbf{Ans.}$$

Ex. 6   A merchant bought goods for £375. At what price must he sell them in order to make a profit of 25% on selling price?

(For S.P. of £100, C.P. will be £75.)

$$\text{S.P.} = £\frac{375 \times 100}{75}$$

$$= £500 \qquad \textbf{Ans.}$$

Ex. 7 By selling goods for £258·75 a dealer makes a profit of 15% on cost price. Find the cost price of the goods.

(For S.P. of £115, C.P. will be £100.)

$$\therefore \qquad \text{C.P.} = £\frac{\overset{9}{\cancel{1035}}}{4} \times \frac{100}{\cancel{115}}$$

$$= £225 \qquad \textbf{Ans.}$$

Ex. 8 By selling a radio for £40 a dealer makes a profit of 17½% on selling price. Find how much the radio cost the dealer.

$$\text{C.P.} = £\frac{\cancel{40} \times \overset{33}{\cancel{82\cdot5}}}{\underset{\cancel{40}}{\cancel{100}}}$$

$$= £33 \qquad \textbf{Ans.}$$

Ex. 9 By selling a house for £25 800 a profit of 7½% was made. At what price should the house have been sold in order to make a profit of 10%?

$$\text{C.P.} = £\frac{25\,800}{1} \times \frac{100}{107\frac{1}{2}}$$

$$\text{S.P.} = £\frac{\overset{600}{\cancel{25\,800}}}{1} \times \frac{100}{\underset{43}{\cancel{107\cdot5}}} \times \frac{\overset{44}{\cancel{110}}}{\cancel{100}}$$

$$= £26\,400 \qquad \textbf{Ans.}$$

Ex. 10 A wholesaler wishes to make a profit of 10% on selling price after allowing the retailer 15% trade discount. What will be his catalogue price for an article which costs him £15·30?

$$£$$

Let catalogue price be	100
Less 15% discount	15
	85
„ 10% profit	8·5
Cost price	76·5

$$\text{Catalogue price} = £\frac{\cancel{15}\cdot3 \times 100}{\underset{5}{\cancel{76\cdot5}}}$$

$$= \pounds 20 \qquad \textbf{Ans.}$$

Ex. 11 A tradesman sold four-fifths of his stock at a profit of 25% and the remainder at a loss of 10%. What was his percentage profit on the whole?

Let C.P. be 100.

$$\text{S.P.} \quad \frac{80}{1} \times \frac{125}{100} = 100$$
$$\frac{20}{1} \times \frac{90}{100} = 18$$
$$\overline{118}$$

Profit on whole = **18%**   **Ans.**

Ex. 12 A tradesman wishes to mark his goods at a price which will give him a profit of $33\frac{1}{3}\%$ on selling price after allowing a cash discount of 5%. At what price will he mark an article which costs him £4·75?

Let marked price be	120
Less 5% discount	6
	114
„ $33\frac{1}{3}\%$ profit	38
Cost price	76

∴ Ratio of cost price to marked price = 76:120 = 19:30.

And marked price = $£\frac{19}{4} \times \frac{30}{19} = \pounds 7\cdot50$ **Ans.**

Exercise 8

1. Goods are bought for £15 and sold for £18·75. Find the percentage profit: (a) on cost price, (b) on selling price.

2. A grocer bought 56 kg of tea for £68 and sold it for £1·60 a kg. Find his percentage profit: (a) on cost price, (b) on selling price. (Answer to one-tenth.)

3. A boy bought a bicycle for £45 and sold it for £49·50. Find his percentage profit.

4. A firm's sales last year were £84 375. The cost of the goods was £58 725. Selling expenses amounted to £17 325. Find: (a) the gross profit, and (b) the net profit expressed as a percentage on turnover.

5. A firm's gross profit on turnover was 27½%, and its net profit on turnover was 12½%. If the selling expenses amounted to £23 925, what was the turnover?

6. The turnover of a firm was £58 500. The gross profit was 37½% and the net profit was 15% on turnover. Find: (a) the selling expenses, and (b) the cost of the goods sold.

7. A car was bought for £950 and sold at a profit of 24%. Find the selling price.

8. A car was bought for £2550 and sold at a loss of 15%. Find the selling price.

9. A merchant bought goods for £425. At what price must he sell them in order to make a profit of 15% on selling price?

10. A grocer bought 144 tins of plums for £40·30. At what price per tin must he sell them in order to make a profit of 25% on selling price?

11. By selling a radio cassette player for £84 a dealer makes a profit of 12% on cost price. Find the cost price.

12. A dealer sells fountain pens at £6·50 each. His profit is 33⅓% on turnover. Find the cost price of 12 of these fountain pens.

13. A dealer sold two-thirds of his stock at a profit of 15% and one-third at a loss of 15%. Find his total percentage profit.

14. By selling a music centre for £211·20 a dealer makes a profit of 10%. At what price should the music centre have been sold in order to make a profit of 17½%?

15. A bought a bicycle for £10 and sold it to B at a profit of 10%. B sold it to C at a loss of 10%. How much did C pay for the bicycle?

16. A man bought 144 tennis balls for £48 and sold them at a profit of 33⅓% on selling price. At what price did he sell each ball?

17. What is the percentage gain or loss in buying oranges at 5 for 30p and selling a half at 7p each and a half at 5½p each?

18. A tradesman wishes to mark his goods at a price that will give him a profit of 33⅓% on selling price after allowing 5% cash discount. At what price will he mark an article which costs him £6·65?

19. A tradesman wishes to mark his goods at a price which will give him a profit of 33⅓% on cost price after allowing a cash discount of 5 per cent. At what price will he mark an article which costs him £2·85?

20. A man buys eggs at a certain price for 20 and sell: them at that price for 12. What is his percentage profits (*a*) on cost price, (*b*) on selling price?

CHAPTER 9

Simple Interest

Interest is a charge for the use of money for a specified period. It is usually reckoned at a certain rate per cent. per annum (*i.e.* for a year). The problem of interest follows naturally upon, and is an extension of, our work in the previous chapter on percentage values. To this is added the element of 'Time'. There are three factors which will determine the amount of the interest. There is the sum of money on which the interest is payable; there is the rate of interest; and there is the time for which it is borrowed. If, for example, we were required to find the interest on £100 *for* 1 *year* at 5% per annum, all we are required to do is to find 5% of £100, which is £5. If, however, the money was required for only 6 months, which is half a year, then the interest would be a half of £5 which is £2·50.

The *Principal* is the sum on which interest is payable.

The *Amount* is the Principal *plus* Interest, *i.e.* the sum to be repaid.

The interest may be for a number of months, in which case we divide the number by 12, or for a number of days, in which case we divide the number by 365.

Simple Interest is interest on a fixed Principal. We mean by this that the interest is non-cumulative as an interest-bearing medium, whatever the length of time. If, for example, we were required to find the interest on £100 for 3 years at 5% per annum, the interest would be 3 times as much as that for 1 year, that is £15. In other words, *the Interest is not added to the Principal until the end of the given time.*

Ex. 1   Find the Simple Interest on £100 for 2 years at 8% per annum.

$$\text{Simple Interest} = \text{\pounds}\frac{100}{1} \times \frac{8}{100} \times \frac{2}{1}$$
$$= \textbf{\pounds16} \qquad \textbf{Ans.}$$

Ex. 2   Find the Simple Interest on £100 for 5 months at 8%.

$$\text{Interest} = \text{\pounds}\frac{100}{1} \times \frac{8}{100} \times \frac{5}{12}$$
$$= \textbf{\pounds3·33} \qquad \textbf{Ans.}$$

Note: Per annum is understood.

Ex. 3   Find the Simple Interest on £100 for 73 days at 8%.

$$\text{Interest} = \text{\pounds}\frac{100}{1} \times \frac{8}{100} \times \frac{73}{365}$$
$$= \textbf{\pounds1·60} \qquad \textbf{Ans.}$$

Note: 73 days = $\frac{1}{5}$ year.

146	„	$= \frac{2}{5}$ „
219	„	$= \frac{3}{5}$ „
292	„	$= \frac{4}{5}$ „

From the above we can construct the following formula:

$$I = \text{\pounds}\frac{P \times R \times Y}{100} \qquad \begin{array}{l} \text{where } I = \text{Interest} \\ P = \text{Principal} \\ Y = \text{Time in years} \\ R = \text{Rate per cent.} \end{array}$$

We can use this formula for the solution of problems in which we are required to find the Principal, Rate, or Time in years.

$$\frac{P \times R \times Y}{100} = I.$$

Multiplying both sides by 100, we get:

$$P \times R \times Y = I \times 100.$$

Dividing both sides by $R \times Y$, we get:

$$(1) \quad P = \frac{I \times 100}{R \times Y}.$$

Dividing both sides by $P \times Y$, we get:

$$(2) \quad R = \frac{I \times 100}{P \times Y}.$$

Dividing both sides by $P \times R$, we get:

$$(3) \quad Y = \frac{I \times 100}{P \times R}.$$

From the above it will be clear that we can find any one of the three factors P, R, Y, when given the interest and two of these factors, by multiplying the interest by 100 and dividing by the two known factors.

$$\text{The unknown factor} = \frac{\text{Interest} \times 100}{\text{Two known factors}}.$$

Ex. 4   Find the Simple Interest on £375 for 7 months at $7\frac{1}{4}\%$ per annum.

$$\text{Interest} = £\frac{375 \times 7\frac{1}{4} \times 7}{100 \times 12}$$

$$= £\frac{\overset{5}{375} \times 29 \times 7}{\underset{16}{400} \quad \underset{4}{12}}$$

$$= \textbf{£15·86} \qquad \textbf{Ans.}$$

$$
\begin{array}{r}
15\cdot859 \\
64)1015(\\
64 \\
\hline
375 \\
320 \\
\hline
550 \\
512 \\
\hline
380 \\
320 \\
\hline
600 \\
576 \\
\hline
24
\end{array}
$$

Note: Interest is always given *to the nearest penny*.

Ex. 5   Find the amount of £126 for 219 days at $7\frac{3}{4}\%$ per annum.

$$\text{Interest} = £\frac{126 \times 7\frac{3}{4} \times 219}{100 \times 365}$$

$$= £\frac{\overset{63}{126} \times 31 \times 3}{\underset{200}{400} \times 5}$$

$$= £5\cdot86.$$

$$\text{Amount} = £126 + £5\cdot86$$

$$= \textbf{£131·86} \qquad \qquad \textbf{Ans.}$$

Ex. 6   What sum will earn £27 interest in 10 months at $9\frac{3}{4}\%$ per annum?

$$\text{Sum} = \text{£}\frac{27 \times 100}{9\frac{3}{4} \times \frac{10}{12}}$$

$$= \text{£}\frac{\overset{9}{\cancel{27}} \times 10\cancel{0} \times 4 \times 12}{\cancel{39} \times \cancel{10}}$$

$$= \text{£332·31}\ ^{13} \qquad\qquad \textbf{Ans.}$$

Ex. 7   At what rate will £650 amount to £689 in 8 months?

$$\text{Interest} = \text{£689} - \text{£650} = \text{£39.}$$

$$\text{Rate} = \frac{39 \times 100}{650 \times \frac{8}{12}}\%$$

$$= \frac{\overset{3}{\cancel{39}} \times \cancel{100} \times \overset{3}{\cancel{12}}}{\cancel{650} \times \cancel{8}}\%$$

$$= \textbf{9\%} \qquad\qquad \textbf{Ans.}$$

Ex. 8   In what time will £840 amount to £886·55 at $9\frac{1}{2}\%$ per annum?

$$\text{Interest} = \text{£46·55.}$$

$$\text{Time} = \frac{931 \times 100}{20 \times 840 \times 9\frac{1}{2}} \text{ year}$$

$$= \frac{\overset{7}{\cancel{931}} \times \cancel{100} \times \cancel{2}}{\cancel{20} \times \underset{12}{\cancel{840}} \times \cancel{19}} \quad \text{,,}$$

$$= \textbf{7 months} \qquad\qquad \textbf{Ans.}$$

Ex. 9   On the 3rd May a man borrowed £292. The debt is to be repaid with interest at $7\frac{1}{2}\%$ per annum on the 3rd August. What will be the amount repaid?

$$\text{Days} = 28 + 30 + 31 + 3 = 92.$$

$$\text{Interest} = \text{£}\frac{\overset{2}{\cancel{292}} \times \overset{3}{\cancel{7}\cdot\cancel{5}} \times 92}{100 \times \cancel{365}}$$

$$= \text{£5·52.}$$

$$\text{Amount} = \pounds 292 + \pounds 5 \cdot 52$$
$$= \pounds 297 \cdot 52 \qquad \textbf{Ans.}$$

Note: In counting days, omit the first and count the last. To do this, subtract the first date from the number of days in that month and then count all the following days.

1. Interest Tables

Interest tables are used to facilitate the work of calculation in banks and business houses where interest calculations are of frequent occurrence. Such tables may be bought. The following is an example of a table which you can construct.

Ex. 10　Construct a table for finding the interest on any amount for 1 day at $7\frac{1}{2}\%$ per annum.

$$\text{Interest on } \pounds 1 \text{ for 1 day} = \pounds\frac{7\frac{1}{2}}{100 \times 365}$$
$$= \pounds\frac{3}{146\,00}$$
$$= \pounds\frac{0 \cdot 03}{146}$$
$$= \pounds 0 \cdot 000\ 205\ 479\ 45$$

£	£	
1	0·000 205 479 45	
2	0·000 410 958 90	
3	0·000 616 438 35	
4	0·000 821 917 80	Interest on £1 to £9 at $7\frac{1}{2}\%$
5	0·001 027 397 25	for 1 day.
6	0·001 232 876 70	
7	0·001 438 356 15	
8	0·001 643 835 60	
9	0·001 849 315 05	

Ex. 11　Find, from the above table, the interest on £97 640 for 37 days.

$$
\begin{array}{rl}
\text{£} & \text{£} \\
90\ 000 = & 18 \cdot 493\ 15 \\
7\ 000 = & 1 \cdot 438\ 35 \\
600 = & 0 \cdot 123\ 28 \\
\underline{40 =} & \underline{0 \cdot 008\ 21} \\
97\ 640 = & 20 \cdot 062\ 99 \\
& \underline{37} \\
& 140 \cdot 440\ 93 \\
& \underline{601 \cdot 889\ 70} \\
& 742 \cdot 330\ 63 = \textbf{£742·33} \qquad \textbf{Ans.}
\end{array}
$$

This table could also be used as a table of interest on £1 for 1 to 9 days at $7\frac{1}{2}\%$. In this case we substitute 'days' for '£' in the first column.

Ex. 12 Find the interest on £15 000 for 358 days at $7\frac{1}{2}\%$ per annum.

$$
\begin{array}{rl}
\text{Days} & \text{£} \\
300 = & 0 \cdot 061\ 643\ 83 \\
50 = & 0 \cdot 010\ 273\ 93 \\
\underline{8 =} & \underline{0 \cdot 001\ 643\ 83} \\
358 = & 0 \cdot 073\ 561\ 63
\end{array}
$$

$$
\begin{array}{rl}
\text{£} & \text{£} \\
10\ 000 = & 735 \cdot 616\ 3 \\
\underline{5\ 000 =} & \underline{367 \cdot 808\ 2} \\
15\ 000 = & 1103 \cdot 424\ 5 = \textbf{£1103·42} \qquad \textbf{Ans.}
\end{array}
$$

2. Third, Tenth, and Tenth Rule

When interest is required for a number of days and neither the Principal nor the days are divisible by 73 it is quicker and easier to double the rate and divide by 73 000.

The interest formula then becomes:

$$
I = \frac{P \times D \times 2R}{73\ 000} \quad \text{where D = number of days.}
$$

$$
\frac{100}{73} = 1 \cdot 37 \text{ (to 2 decimal places).}
$$

∴ To divide by 73:   Divide by 100 and multiply by 1·37.

      ,,   ,,   73 000:   Divide by 100 000 and multiply by 1·37.

$1·37 = 1\frac{37}{100} = 1\frac{111}{300}$

$\quad\quad = 1 + \frac{1}{3} + \frac{1}{30}$ (*i.e.* $\frac{1}{10}$ of $\frac{1}{3}$) $+ \frac{1}{300}$ (*i.e.* $\frac{1}{10}$ of $\frac{1}{10}$ of $\frac{1}{3}$).

Rules for dividing by 73 000:

  (*a*) Divide by 100 000 (move the decimal point 5 places to the left).

  (*b*)   ,,   (*a*) by 3
  (*c*)   ,,   (*b*) ,, 10
  (*d*)   ,,   (*c*) ,, 10
  (*e*) Add the above.

Ex. 13   Find the amount of £750 invested for 128 days at $6\frac{1}{4}\%$ simple interest.

$$\text{Interest} = £\frac{750 \times 128 \times 12\frac{1}{2}}{73\,000} \quad \begin{bmatrix} 12·000\,00 \\ 4·0 \\ 0·4 \\ 0·04 \\ \hline 16·44 \end{bmatrix}$$

$$= £16·44.$$

$$\text{Amount} = \textbf{£766·44} \quad\quad \textbf{Ans.}$$

Note: $\frac{100}{73} = 1·3699$ (to 4 decimal places)

$\quad\quad 1·37 - 1·3699 = 0·0001.$

In some cases, therefore, it may be necessary to correct the interest obtained by the above method by deducting $\frac{1}{10000}$ part or 1p per £100.

$$\begin{aligned} e.g. \text{ Interest} \quad &= £17·8862 \\ \text{Less } \tfrac{1}{10000} = \quad &\underline{0·0017} \\ &£17·8845 = \textbf{£17·88} \end{aligned}$$

Ex. 14   Check, by using the Third, Tenth, and Tenth Rule, the answer to Ex. 11 above.

$$\text{Interest} = \pounds\frac{97\ 640 \times 37 \times 15}{73\ 000}$$

$$= \pounds742\cdot33$$

∴   Answer correct.

$$\begin{bmatrix} 14\cdot646\ 00 \\ \underline{37} \\ 102\cdot522\ 00 \\ 439\cdot380\ 00 \\ \hline 541\cdot902\ 00 \\ 180\cdot634\ 00 \\ 18\cdot063\ 40 \\ \underline{1\cdot806\ 34} \\ 742\cdot405\ 74 \\ \underline{\cdot074\ 24} \\ 742\cdot331\ 50 \end{bmatrix}$$

Ex. 15   £12 000 was placed on deposit at 11½% and withdrawn with interest 75 days later. Find the amount withdrawn.

$$\text{Interest} = \frac{\pounds12\ 000 \times 75 \times 23}{73\ 000}$$

$$= \pounds283\cdot59 - 3p$$
$$= \pounds283\cdot56$$

Amount = **£12 283·56**   **Ans.**

$$\begin{bmatrix} 198\cdot000\ 00 \\ 9\cdot000\ 00 \\ \hline 207\cdot000\ 00 \\ 69\cdot000\ 00 \\ 6\cdot900\ 00 \\ 0\cdot690\ 00 \\ \hline 283\cdot590 \end{bmatrix}$$

Exercise 9

Find the simple interest on:

1. £245 for 73 days   at 7½% per annum.
2. £540 ,, 146 ,,   ,, 7½%   ,,
3. £340 ,, 219 ,,   ,, 7½%   ,,
4. £450 ,, 292 ,,   ,, 7½%   ,,
5. £650 ,, 4 years   ,, 6¾%   ,,
6. £438 ,, 55 days   ,, 9%   ,,
7. £400 ,, 7 months ,, 10½%   ,,
8. £360 ,, 11 ,,   ,, 11¼%   ,,
9. £470 ,, 9 ,,   ,, 9%   ,,
10. £360 ,, 165 days   ,, 9%   ,,

Find the amount of:

11. £280 for   8 months at   6¾% per annum.
12. £657   ,,   89 days   ,,   7½%   ,,
13. £803 ,, 147   ,,   ,,   11¼%   ,,
14. £375 ,, 146   ,,   ,,   12¾%   ,,
15. £876 ,,   75   ,,   ,,   9%   ,,

What sum will earn:

16. £7·50 interest in   8 months at   7½% per annum?
17. £9   ,,   219 days   ,,   11¼%   ,,
18. £24·75   ,,   11 months ,,   7¼%   ,,
19. £2·80   ,,   73 days   ,,   7½%   ,,
20. £18·20   ,,   7 months ,,   9¾%   ,,

At what rate will:

21. £300 earn £33   interest in   11 months?
22. £200   ,,   £6   ,,   146 days?
23. £180   ,,   £7·50   ,,   5 months?
24. £219   ,,   £1·35   ,,   50 days?
25. £240   ,, £15·30   ,,   9 months?

In what time will:

26. £360 amount to £370·50 at   5% per annum?
27. £500   ,,   £520   ,, 10%   ,,
28. £480   ,,   £492   ,,   7½%   ,,
29. £511   ,,   £523·60 ,,   4½%   ,,
30. £450   ,,   £480   ,,   8%   ,,

Construct a table for finding the interest on £1 to £9 for 1 day at 4% per annum.
From your table find the interest on:

31. £1475 for 35 days at 12% per annum.
32. £2389 ,,   74   ,,   12%   ,,
33. £876 ,, 219   ,,   12%   ,,
34. £876 ,, 150   ,,   12%   ,,

35. Find the simple interest on £730 from the 15th January, 1972, to the 14th June, 1972, at 10½% per annum.
36. A man borrowed £280 on the 5th March, and on the

10th April he borrowed another £220. On the 1st July he repaid the loan with interest at 6% per annum. How much was repaid?

37. On the 1st March a man borrows £500 at 10% per annum, and agrees to repay the loan and interest when the interest is exactly £20. On what date will he repay the loan?

38. A man borrowed a certain sum at 11¼% per annum, and 6 months later repaid the Principal and Interest, which amounted to £845. What was the sum borrowed?

39. A man's account at the bank was overdrawn to the extent of £85 from the 10th March to the 14th November. If the bank charges 13% interest, what will the interest amount to on the 14th November?

40. In what time, to the nearest month, will a sum of money double itself at 10½% per annum simple interest?

41. Find the simple interest on £550 invested for 160 days at 5¾% per annum.

42. £25 000 was borrowed for 48 days at 7¼% simple interest. Find the amount to be repaid.

CHAPTER 10

Compound Interest—Depreciation

1. Compound Interest

In *Compound Interest*, the Principal does not remain a fixed sum as in simple interest, but is cumulative. At the end of each stated period the Interest is added to the Principal and for the next and subsequent periods the Interest is calculated on the Principal *plus* Interest of the previous period. This will be much clearer if the following examples are carefully studied. In compound interest we first find the amount and then deduct the Principal to find the Interest.

Ex. 1   Find the compound interest on £496 for 3 years at 5% per annum.

> *1st method.* Express the rate per cent. as a decimal and add 1. Multiply the principal by this number for the 1st year. Multiply the amount thus obtained by this same number for the 2nd year and so on for each succeeding year.

$5\% = 0.05$,   ∴ the multiplier is 1.05.

$$
\begin{array}{r}
£ \\
496 \\
1.05 \\
\hline
496 \\
24.8 \\
\hline
520.8 \\
1.05 \\
\hline
520.8 \\
\end{array}
\quad \text{1st year}
$$

continued

$$\frac{26\cdot04}{546\cdot84} \quad \text{2nd year}$$

$$\frac{1\cdot05}{546\cdot84}$$
$$\frac{27\cdot342}{574\cdot182} \quad \text{3rd year}$$

Interest = £574·18 — £496 = **£78·18**   **Ans.**

2nd method. This method is preferable when the rate can be expressed as a convenient fraction. In this method we divide each Principal by the denominator of the fraction, and add the result to the Principal for the new amount.

$$5\% = \tfrac{1}{20} = \begin{array}{l} £ \\ 496 \\ \underline{24\cdot8} \\ 520\cdot8 \quad \text{1st year} \\ \underline{26\cdot04} \\ 546\cdot84 \quad \text{2nd year} \\ \underline{27\cdot342} \\ 574\cdot182 \quad \text{3rd year} \end{array}$$

Interest = £574·18 — £496 = **£78·18**   **Ans.**

3rd method. This method may be used with advantage where the rate per cent. is an integral number of not more than 12. It is a shortened version of the 1st method.

Multiply the Principal each year by the rate, and put the product two places to the right, then add to the Principal.

Multiply by 5 and put
the product 2 places to
the right

$$\begin{array}{l} £ \\ 496 \\ \underline{24\cdot8} \\ 520\cdot8 \quad \text{1st year} \\ \underline{26\cdot04} \\ 546\cdot84 \quad \text{2nd year} \quad \textit{continued} \end{array}$$

$$\frac{27 \cdot 342}{574 \cdot 182} \quad \text{3rd year}$$

Interest $= £574 \cdot 18 - £496 = $ **£78·18**   **Ans.**

Ex. 2   Find the amount of £275 for 3 years at 4% per annum compound interest.

$$£$$

$$
\begin{array}{ll}
 & 275 \\
4\% = & \underline{11} \\
 & 286 & \text{1st year} \\
4\% = & \underline{11 \cdot 44} \\
 & 297 \cdot 44 & \text{2nd year} \\
4\% = & \underline{11 \cdot 8976} \\
 & 309 \cdot 3376 & \text{3rd year}
\end{array}
$$

Amount $= $ **£309·34**   **Ans.**

Ex. 3   Find the compound interest on £400 for 3 years at 7% per annum.

$$£$$

$$
\begin{array}{ll}
 & 400 \\
7\% = & \underline{28} \\
 & 428 & \text{1st year} \\
7\% = & \underline{29 \cdot 96} \\
 & 457 \cdot 96 & \text{2nd year} \\
7\% = & \underline{32 \cdot 0572} \\
 & 490 \cdot 0172 & \text{3rd year}
\end{array}
$$

Interest $= £490 \cdot 02 - £400 = $ **£90·02**   **Ans.**

When, in compound interest, the interest is to be added at the end of periods less than a year, *i.e.* monthly, quarterly, or half-yearly, the rate is divided and a correspondingly greater number of periods is taken.

Ex. 4   Find the compound interest on £480 for 2 years at 5% per annum, interest to be added half-yearly.

$$2\tfrac{1}{2}\% = \tfrac{1}{40} = \begin{array}{r} \pounds \\ 480 \\ \underline{12} \\ 492 \\ \underline{12\cdot3} \\ 504\cdot3 \\ \underline{12\cdot6075} \\ 516\cdot9075 \\ \underline{12\cdot9227} \\ 529\cdot8302 \end{array}$$

$\frac{1}{2}$ year (492)

1 year (504·3)

$1\frac{1}{2}$ years (516·9075)

2 years (529·8302)

Interest = £529·83 − £480 = **£49·83**   **Ans.**

Ex. 5   Find the difference between the simple and compound interest on £375 for 3 years at 3% per annum.

$$\text{Simple Interest} = \pounds\frac{375 \times 3 \times 3}{100}$$

$$= \pounds 33\cdot75$$

Compound Interest

$$\begin{array}{r} \pounds \\ 375 \\ 3\% = \underline{11\cdot25} \\ 386\cdot25 \\ 3\% = \underline{11\cdot5875} \\ 397\cdot8375 \\ 3\% = \underline{11\cdot9351} \\ 409\cdot7726 \end{array}$$

1st year (386·25)

2nd year (397·8375)

3rd year (409·7726)

Interest = £409·77 − £375 = £34·77.

Difference = £34·77 − £33·75

$= $ **£1·02**   **Ans.**

Ex. 6   Find the amount of £440 for 2 years at $3\frac{1}{2}\%$ per annum compound interest.

$$\begin{array}{r} \pounds \\ 440 \\ \underline{1\cdot035} \\ 440 \end{array}$$

continued

$$13\cdot2$$
$$\underline{2\cdot2}$$
$$\overline{455\cdot4}\qquad\text{1st year}$$
$$13\cdot662$$
$$\underline{2\cdot277}$$
$$\overline{471\cdot339}\qquad\text{2nd year}$$

Amount = **£471·34** **Ans.**

Note: There is no need to write down the multiplier more than once. The product can then serve as the first multiplication by 1.

Another method:

$$£$$
$$440$$
$$2\tfrac{1}{2}\% = \tfrac{1}{40} = \quad 11$$
$$1\% \ = \tfrac{1}{100} = \underline{\quad 4\cdot4}$$
$$\overline{455\cdot4}\qquad\text{1st year}$$
$$11\cdot385$$
$$\underline{4\cdot554}$$
$$\overline{471\cdot339}\qquad\text{2nd year}$$

Amount = **£471·34** **Ans.**

Ex. 7 A man wishes to invest a sum of money at 6% per annum compound interest which will amount to £5000 in 3 years. What sum must he invest?

$$£$$
Let sum be 100
$$6\% = \quad 6$$
$$\overline{106}\qquad\text{1st year}$$
$$6\% = \quad 6\cdot36$$
$$\overline{112\cdot36}\qquad\text{2nd year}$$
$$6\% = \quad 6\cdot7416$$
$$\overline{119\cdot1016}\qquad\text{3rd year}$$

For £100 invested now he will receive £119·1016 in 3 years.

∴ The amount he must invest $= £\dfrac{5000 \times 100}{119{\cdot}1016}$.

$$
\begin{array}{r}
4198 \\
119{\cdot}1016)\overline{500000}(\\
476406 \\
\hline
23594 \\
11910 \\
\hline
11684 \\
10719 \\
\hline
965 \\
953 \\
\hline
8
\end{array}
$$

To nearest £1 $= $ **£4198**   **Ans.**

2. Depreciation

Depreciation is a loss in value. The fixed assets of a business, such as buildings, furniture and fixtures, plant, and machinery, lose in value with the passage of time, and in some cases become obsolete. In order to show their true value, it is customary to write down their value periodically. The object of this is to show the assets at a figure approximating to their market value and to prevent an over-statement of profits.

The fixed percentage method of depreciation is the writing down of the book value of an asset by a fixed percentage of the residual value each year.

Ex. 8   A manufacturer buys a machine for £6000 and writes off 10% at the end of each year. Find its value at the end of 4 years.

$$
\begin{array}{r}
£ \\
6000 \\
10\% = \tfrac{1}{10} = \quad 600 \\
\hline
5400 \qquad \text{1st year} \\
540 \\
\hline
4860 \qquad \text{2nd year}
\end{array}
$$

continued

$$\begin{array}{r} 486 \\ \hline 4374 \\ 437{\cdot}4 \\ \hline 3936{\cdot}6 \end{array}$$ 3rd year

4th year

Value = **£3936·60**   **Ans.**

3. Other Problems

Ex. 9   In 1941 a city had a population of 502 157. It is estimated that it is increasing at the rate of 120 per thousand every 10 years. Find its estimated population in 1981.

(120 per thousand = 12%.)

$$12\% = \begin{array}{r} 502\ 157 \\ 60\ 258 \\ \hline 562\ 415 \\ 67\ 489 \\ \hline 629\ 904 \\ 75\ 588 \\ \hline 705\ 492 \\ 84\ 659 \\ \hline 790\ 151 \end{array}$$

1941
(whole numbers only)

1951

1961

1971

1981

Estimated population = **790 151**   **Ans.**

When a person has a sum of money invested for a long period in a particular venture or property, he may take into consideration its interest-bearing capacity when considering the success or failure of the investment and the true rate of profit or loss.

Ex. 10   A merchant bought a diamond necklace for £4000. He sold it 3 years later for £5000. In reckoning his profit on the sale he takes into consideration the fact that his money could have been invested for this time at 3% per annum compound interest. Find his true rate of profit.

$$£$$
$$4000$$
$$3\% = \underline{120}$$
$$4120 \qquad \text{1st year}$$
$$\underline{123\cdot6}$$
$$4243\cdot6 \qquad \text{2nd year}$$
$$\underline{127\cdot308}$$
$$4370\cdot908 \qquad \text{3rd year} = \text{approx. } £4370\cdot9.$$

Profit $= £5000 - £4370\cdot9 = £629\cdot1$.

$$\text{Rate} = \frac{629\cdot1 \times 100\%}{4370\cdot9}$$

$$= \textbf{14·4\%} \qquad \textbf{Ans.}$$

$$\begin{array}{r} 14\cdot4 \\ 43709)\overline{629100(} \\ \underline{43709} \\ 192010 \\ \underline{174836} \\ 17174 \end{array}$$

Ex. 11   Find the present value of £800 due in 2 years' time, reckoning compound interest at 6% per annum.

£1 at 6% per annum for 2 years $= £1\cdot1236$
The P.V. of £1·1236 $= £1$

$\therefore$ the   ,,   £800 $= £\dfrac{800}{1\cdot1236} = \textbf{£712}$ **Ans.**

Note: P.V. $= \dfrac{A}{1\cdot0r^n}$.

Ex. 12   A machine was bought 3 years ago, and 10% has been written off its value at the end of each year. Its book value is now £3645. Find its cost price.

$$\text{C.P.} = £\frac{3645}{(1 - 0\cdot1)^3} = £\frac{3645}{0\cdot9^3} = £\frac{3645}{0\cdot729} = \textbf{£5000} \quad \textbf{Ans.}$$

Note: $0 = \dfrac{\text{P.V.}}{(1 - 0\cdot0r)^n}$ where $0 =$ original value.

Exercise 10

Find the amount of:

1. £480 in 3 years at $2\frac{1}{2}\%$ compound interest.
2. £640 ,, 3 ,, 5% ,,
3. £740 ,, 4 ,, 4% ,,
4. £560 ,, 3 ,, $3\frac{1}{2}\%$,,
5. £450 ,, 2 ,, $3\frac{3}{4}\%$,,
6. £200 ,, 3 ,, $4\frac{1}{2}\%$,,

Find the compound interest on:

7. £800 in 2 years at $2\frac{3}{4}\%$ compound interest.
8. £640 ,, 3 ,, $3\frac{1}{2}\%$,,
9. £364 ,, 3 ,, $2\frac{1}{2}\%$,,
10. £900 ,, 3 ,, 4% ,,
11. £280 ,, 3 ,, $4\frac{1}{2}\%$,,
12. £600 ,, 4 ,, 5% ,,

13. Find the compound interest on £100 for 3 years at 5% per annum.

14. A owes B £600 due in 3 years' time. How much should B receive now in discharge of the debt, assuming that he can obtain 5% compound interest on his money?

15. Find the compound interest on £250 for 2 years at 4% per annum, interest added: (*a*) yearly, (*b*) half-yearly.

16. The population of a town was 35 000 in 1938. If it increased at the rate of 20% every ten years, what was the population in 1968?

17. Find the compound interest on £350 for 2 years at 6% per annum, interest being added half-yearly.

18. Find the difference between the simple and compound interest on £450 for 3 years at 4% per annum.

19. What sum, to the nearest 10p, must a man invest now in order to receive £600 in 4 years' time, reckoning compound interest at 4% per annum?

20. A manufacturer buys a machine for £800 and writes off 12½% for depreciation at the end of each year. What will its book value be at the end of 3 years?

21. A merchant bought an antique table for £450 and 3 years later sold it for £550. Allowing that he could have received 4% compound interest on his money, what was his true rate of profit?

22. A merchant writes off 10% for depreciation of his plant and machinery. Its present book value is £5382. What was its value 3 years ago?

23. A firm writes off 7½% for depreciation of fixtures and fittings. They are at present valued at £536. What will be their book value 2 years hence?

24. A company writes off 5% per annum for depreciation of land and buildings. Their present value is £37 560. What, to the nearest £1, will be their book value in 4 years' time?

CHAPTER II

Ratio and Proportion

Ratio is the relationship with regard to magnitude which exists between quantities *of the same kind*. The relationship which exists between 3 tonnes and 7 tonnes may be expressed in the form 3:7, which is stated 'as 3 is to 7'. This is a corruption of $3 \div 7$, and a ratio may be expressed in fractional form, thus, $3:7 = \frac{3}{7}$.

A ratio can only exist between quantities of *exactly the same kind*, but we can obtain a ratio between quantities of weight or capacity or length or area or volume or money, etc., by bringing them to a common unit of comparison. Thus, the ratio between:

$$240 \text{ g and } 3 \text{ kg} \quad = \quad 240:3000 = \tfrac{2}{25}.$$
$$50 \text{ a and } 2 \text{ ha} \quad = \quad 50:200 \ = \tfrac{1}{4}.$$
$$750 \text{ m and } 1 \text{ km} \quad = \quad 750:1000 = \tfrac{3}{4}.$$
$$45 \text{ mm}^2 \text{ and } 1 \text{ cm}^2 = \quad 45:100 \ = \tfrac{9}{20}.$$
$$80\text{p and } \pounds 2 \quad = \quad 80:200 \ = \tfrac{2}{5}.$$

It will be obvious from the above that the terms of a ratio may be reduced by division in exactly the same way as we reduce a fraction by cancellation. In fact a ratio is a fraction, and when expressed as a fraction it denotes the relationship of the 1st quantity to the 2nd quantity when the 2nd quantity is considered as unity, or the unit of comparison.

Ex. 1   Give the ratio between 60p and £2.

$$\text{Ratio} = 60:200$$
$$= \quad 3:10$$
$$= \tfrac{3}{10}:1$$

or just $\tfrac{3}{10}$, 1 being understood.

i.e. 60p is $\tfrac{3}{10}$ of a unit of £2.

Ex. 2 Find the ratio of the cost price to the selling price of an article which will give a profit of 25% on cost price.

$$\text{Ratio} = 100:125$$
$$= \mathbf{4:5} \qquad \textbf{Ans.}$$

Ex. 3 Find the ratio of the cost price to the selling price of an article which will give a profit of 25% on selling price.

$$\text{Ratio} = 75:100$$
$$= \mathbf{3:4} \qquad \textbf{Ans.}$$

Ex. 4 Find the ratio between the selling price which will give a profit of 25% on the cost price and the selling price which will give a profit of 25% on the selling price, the cost price being the same in each case.

Let cost price be £100.

$$\text{Then, 1st case, ratio} = 125:100 = \frac{125}{100}$$

$$\text{2nd case, ratio} = 133\tfrac{1}{3}:100 = \frac{133\tfrac{1}{3}}{100}$$

$$\text{and required ratio} = \frac{125}{100}:\frac{133\tfrac{1}{3}}{100}$$
$$= 125:133\tfrac{1}{3}$$
$$= 375:400$$
$$= \mathbf{15:16} \qquad \textbf{Ans.}$$

Ex. 5 The composition by weight of an alloy is 5 parts copper, 2 parts antimony, and 7 parts tin. What weight of each constituent is there in 112 kg of the alloy?

$$\text{Units of weight} = 5:2:7 = 14.$$

Weight of copper $= \tfrac{5}{14}$ of 112 kg $= \mathbf{40\ kg}$ ⎫
   ,,   antimony $= \tfrac{2}{14}$,, $= \mathbf{16\ kg}$ ⎬ **Ans.**
   ,,   tin $= \tfrac{7}{14}$,, $= \mathbf{56\ kg}$ ⎭

2. Proportion

Proportion is the equality of ratios. In other words, when two ratios are equal they are said to be in proportion.

Thus, 2:4 and 5:10 are equal ratios and would be written down as follows:

$$2:4::5:10$$

which is stated 'as 2 is to 4 so is 5 to 10'.

We have four terms and they are called the 1st, 2nd, 3rd, and 4th terms.

means

$$2:4::5:10$$

extremes

The 1st and 4th terms are called the *extremes*.
The 2nd ,, 3rd   ,,   ,,   *means*.
The product of the means = the product of the extremes.

Thus, $4 \times 5 = 2 \times 10$.

3. Inverse Proportion

The *inverse* of a ratio is obtained by inverting the fraction representing the ratio.

The inverse of $\frac{4}{5}$ is $\frac{5}{4}$.

When two quantities are so connected that any change in the ratio of the first is accompanied by a change in an opposite direction of the second, then the second is said to vary inversely, or to be in inverse proportion to the first.

The number of men employed and the time taken to do a piece of work are inversely proportional.

Ex. 6   If 8 men take 12 hours to do a piece of work, how long should 6 men take?

Ratio of men $= \frac{6}{8}$.
Ratio of time taken $= \frac{8}{6}$ of 12 h.
Time $= \textbf{16 h}$ **Ans.**

Consider the following:

Men	4	6	8	12
Time in hours	24	16	12	8
Men time	96	96	96	96

In the above two series of numbers it will be observed that the product of any two corresponding numbers is always the same. We may, therefore, state that one number varies inversely to another when the product of a series of the two numbers is *constant*.

The constant can be determined by multiplication of the two terms and then used to determine the missing term.

Ex. 7   If 10 men can do a piece of work in 18 hours, how long would 9 men take?

$$(\text{Constant} = 10 \times 18 = 180.)$$

∴   9 men take $\frac{180}{9} = $ **20 h**   **Ans.**

Ex. 8   If 10 men can do a piece of work in 18 hours, how many men would be required to do the work in 15 hours?

Number required $= \frac{180}{15} = $ **12 men**   **Ans.**

Ex. 9   The volume of a gas varies inversely as the pressure. The volume is 4 dm³ when the pressure is 45 kg/cm². What will be: (*a*) the pressure when the volume is 5 dm³, and (*b*) the volume when the pressure is 60 kg/cm²?

$$(\text{Constant} = 45 \times 4 = 180.)$$

∴   Pressure $= \frac{180}{5} = $ **36 kg**   **Ans.** (*a*)

Volume $= \frac{180}{60} = $ **3 dm³**   **Ans.** (*b*)

The following are three methods of solving proportion problems:

Ex. 10   If 12 kg lard cost £5·88, what will be the cost of 5 kg?

(*a*) *The Unitary Method*

Cost of 12 kg $= 588$p

,,   1 ,, $= \dfrac{588}{12}$p

,,   5 ,, $= \dfrac{588 \times 5}{12}$p

$= 245$p

$= £2\cdot45$ **Ans.**

(*b*) *The Rule of Three*

Let $x =$ cost required in new pence.
Then $12:5::588:x$.

$$x = \frac{5 \times 588}{12}\text{p}$$

$= 245$p

and cost $= £2\cdot45$ **Ans.**

The following points should be noted:

- (i) The product of the means = the product of the extremes, so that when given 3 terms it is quite easy to find the 4th.
- (ii) Place the denomination of the answer (*i.e.* new pence) in the 3rd term.
- (iii) Decide whether the answer will be larger or smaller than the 3rd term.
- (iv) If *larger*, place the larger of the two remaining numbers in the 2nd term. If *smaller*, place the smaller of the two remaining numbers in the 2nd term. The remaining number is placed in the 1st term.
- (v) The answer will be the product of the *means* divided by the extreme in the 1st term, the denomination being that of the 3rd term.

(c) *The Fractional Method*

$$\text{Cost required} = \frac{588}{1} \times \frac{5}{12}\text{p}$$
$$= \text{£2·45} \qquad \textbf{Ans.}$$

(i) The denomination of the answer is written down in fractional form.

(ii) It is then multiplied by the ratio, which is also written down in fractional form.

(iii) If the answer is going to be *larger*, the ratio will be an improper fraction; if *smaller*, it will be a proper fraction.

This is the quickest method, and the one which the student is advised to practise.

Ex. 11   If 6 men reap a farm in 15 days, working 12 hours a day, how long will 9 men take, working 10 hours a day?

(Ratio of men—shorter time.)
(Ratio of hours—longer time.)
Time taken = $\frac{15}{1} \times \frac{6}{9} \times \frac{12}{10}$ days.
$$= \textbf{12 days} \qquad \textbf{Ans.}$$

Ex. 12   If 6 men, working 10 hours a day, dig a trench 500 m long, 5 m deep, and 4 m wide in 8 days, how long will it take 8 men, working 9 hours a day, to dig a trench 350 m long, 6 m deep, and 5 m wide, if each man in the latter gang can do as much in 1 hour as each man in the former gang takes 1 hour 10 minutes to do?

$$\text{Time} = \frac{8}{1} \times \frac{6}{8} \times \frac{10}{9} \times \frac{60}{\underset{z}{70}} \times \frac{350 \times 6 \times 5}{500 \times 5 \times 4} \text{ days}$$

$$= \textbf{6 days} \qquad \textbf{Ans.}$$

Ex. 13   If the price of diamonds per carat varies as their weight and a diamond of 7 carats is worth £2450, find the value of a diamond of 9 carats.

$$\text{Value} = \pounds\frac{2450}{1} \times \frac{9}{7} \times \frac{9}{7}$$
$$= \pounds 4050 \qquad \text{Ans.}$$

The operative words are *price per carat*.
Dissecting the solution we get:

(1) Price per carat for a 7 carat diamond $= \pounds\dfrac{2450}{7} = \pounds 350$

(2) Price per carat for a 9 carat diamond $= \pounds\dfrac{350}{1} \times \dfrac{9}{7} = \pounds 45$

$\therefore$ (3) Price of a 9 carat diamond $= \pounds 450 \times 9 = \pounds 4050$ **Ans.**

Exercise 11

1. The perimeter of a triangular field is 440 metres, and the ratio of the sides is 5:8:9. Find the length of the sides.

2. Find the ratio between the cost price and the selling price of an article which will give a profit of 10% on cost price.

3. Find the ratio between the cost price and the selling price of an article which will give a profit of 10% on selling price.

4. Find the ratio between the selling price which will give a profit of 20% on cost price and the selling price which will give a profit of 20% on selling price, the cost price being the same in each case.

5. Sand and cement are mixed in the ratio of 3:1 to make up 112 kg. How much sand must be added to make the ratio 6:1?

6. Express in its simplest form the ratio of 1·55 kg to 2·8 kg?

7. If 7 m of carpet cost £62·93, how much will 36 m cost?

8. A train travels 39 km in 45 minutes. How far, at the same average speed, will it travel in 2 hours 20 minutes?

9. If $25\frac{1}{2}$ tonnes of provisions are sufficient for 1700 men for 6 weeks, how much will be required for 2400 men for 4 weeks?

10. A train travels 90 km in $2\frac{1}{2}$ hours. How long, at the same average speed, will it take to complete a journey of 144 km?

11. A bonus of £140 is divided between 3 men in the ratio of their weekly wages. If these are £84, £96, and £100, how much will they each receive?

12. How much tin must be added to 112 kg of copper to make an alloy consisting of 8 parts copper to 3 parts tin?

13. If the painting of a wall 35 m by 13 m costs £36·40, what will be the cost of painting a wall 26 m by 14 m? (Answer to the nearest penny.)

14. If 15 ha of turnips will feed 100 sheep for 30 days, how many ha of turnips will be required to feed 60 sheep for 40 days?

15. If 42 men reap a farm in 11 days, working 10 hours a day, how long will 35 men take, working 11 hours a day?

16. If 728 kg cost £8·40, what will be the cost of 39 kg?

17. The volume of gas varies inversely as the pressure. If the volume is 80 dm³ when the pressure is 6 kg/cm², give: (*a*) the constant for finding a missing term, (*b*) the volume when the pressure is 8 kg/cm², (*c*) the pressure when the volume is 120 dm³.

18. A building contractor estimates that 4 men can build 2 houses in 15 weeks. The work is held up for 3 weeks by bad weather. How many men will be required to finish the work on time?

19. 3 tonnes of coal are bought for £255 and sold at a profit of 25%. Find the selling price of: (a) 250 kg, (b) 950 kg, (c) 700 kg.

20. If 5 men take 42 hours to complete a job, how long should 7 men take?

CHAPTER 12

Proportional Parts—Partnerships

1. Proportional Parts

Suppose the sum of £30 is to be divided between two persons, one person receiving 4 times as much as the other. The ratio of their shares is 4:1. If we divide the whole into 5 parts, one will receive 4 parts and the other 1 part.

$$\text{The amount} = \tfrac{4}{5} \text{ of } £30 = \textbf{£24}$$
$$\text{and } \tfrac{1}{5} \quad \text{,,} \quad = \textbf{£6}$$

Ex. 1   A, B, and C share between them the sum of £4·40. For every 2p A receives, B receives 4p, and C 5p. How much do they each receive?

(Total units = 2 + 4 + 5 = 11.)
A receives $\tfrac{2}{11}$ of £4·40 = **80p** ⎫
B ,, $\tfrac{4}{11}$,, = **£1·60** ⎬ **Ans.**
C ,, $\tfrac{5}{11}$,, = **£2** ⎭

Ex. 2   A, B, and C share a sum of money in the ratio of 4:5:6 respectively. If C received £20, what was the total sum?

$$\text{C's share} = \tfrac{6}{15} = \tfrac{2}{5} = £20.$$
$$\text{Total sum} = £\frac{20}{1} \times \frac{5}{2}$$
$$= \textbf{£50} \qquad \textbf{Ans.}$$

Ex. 3   A man left ½ his property to his wife, ¼ of the remainder to his son, and the rest to be divided equally between his 3 daughters. If each daughter received £5000, what was the total amount of his property?

Wife receives $\frac{1}{2}$.　　　　　　Remainder $= \frac{1}{2}$.

Son　　,,　$\frac{1}{4}$ of $\frac{1}{2} = \frac{1}{8}$.　　　　,,　　$= \frac{1}{2} - \frac{1}{8} = \frac{3}{8}$.

Each daughter $\frac{1}{8} = $ £5000

$\qquad$ Total $= $ £5000 $\times 8 = $ **£40 000** 　　　**Ans.**

2. Partnerships

A *partnership* consists of two or more persons who are carrying on a business with a view to profit.

A Deed of Partnership is a document setting forth the conditions by which the partnership will be governed. Any dispute that may arise will be settled by its terms. It will deal with such matters as the capital of each partner, profits and losses and how they are to be shared, partners' salaries, interest on capital, drawings and loans, the duration of the partnership, etc.

If there is no agreement to the contrary, profits and losses are shared equally.

In some cases, particularly where no arrangement is made for interest on capital, profits may be shared in proportion to capital. This method is very common in textbooks on book-keeping and arithmetic and gives rise to some very interesting problems in proportional parts.

Ex. 4　　The profits of a business are to be shared in proportion to capital invested. A's capital is £45 000 and B's is £35 000. If the profits this year are £12 400, how much will they each receive?

$\qquad$ (Ratio $= 4\frac{1}{2}:3\frac{1}{2} = 9:7$.)

$\qquad$ A's share $= \frac{9}{16}$ of £12 400 $= $ **£6975**$\Big\}$ 　**Ans.**

$\qquad$ B's　,,　$= \frac{7}{16}$　,,　　$= $ **£5425**

Ex. 5　　The profits of a business are to be divided as follows: A is to receive a salary of £500 and the rest of the profits are to be divided in proportion to capital. A's capital is £3000 and B's capital is £5000. The profits are £2740. How much will they each receive?

Divisible profit = £2740 − £500 = £2240.

A receives $\frac{3}{8}$ of £2240 + £500 = **£1340** ⎱

B   ,, $\frac{5}{8}$,,   = **£1400** ⎰   **Ans.**

Ex. 6   A contributed £400, B £600, and C £500 to a joint commercial venture. The profit or losses are to be shared in proportion to capital invested. The venture showed a net profit of £375. How should this amount be distributed?

Total capital = £1500.
Profit   = £375.
            = 25p for each £1 subscribed or $\frac{1}{4}$ of the capital.

A's share = **£100** ⎱

B's ,,   = **£150**   **Ans.**

C's ,,   = **£125** ⎰

When profit is divided in proportion to capital, it may be necessary to take into consideration both the amount of the capital invested and the time for which it has been invested.

Ex. 7   A commences a business with a capital of £3000, 3 months later B enters the business with a capital of £4000, and 2 months after this C is taken into the partnership with a capital of £2500. The profits are to be divided in proportion to the capital invested. At the end of 12 months the profits amount to £716. How much should they each receive?

A. £3000 for 12 months = £36 000 for 1 month.
B. £4000 ,,   9   ,,   = £36 000   ,,   ,,
C. £2500 ,,   7   ,,   = £17 500   ,,   ,,
              Total = £89 500.

A's share = $\frac{72}{179}$ of £716 = **£288** ⎱

B's ,,   = ,,   ,,   = **£288**   **Ans.**

C's ,,   = $\frac{35}{179}$,,   = **£140** ⎰

Ex. 8   A and B are in partnership with capitals of
£5000 and £8000 respectively. Each partner is to receive
from the profits 5% interest on capital and the rest of the
profits are to be divided equally. How much will they each
receive from a profit of £3318?

$$
\left.
\begin{array}{l}
\text{A's interest} = £\ 250 \\
\quad\text{,, profit}\quad = £1334 = \textbf{£1584} \\[4pt]
\text{B's interest} = £\ 400 \\
\quad\text{,, profit}\quad = £1334 = \textbf{£1734}
\end{array}
\right\} \quad \textbf{Ans.}
$$

Exercise 12

1. A and B share £16. A receives 4 times as much as B.
How much does B receive?

2. The rent and rates of a house amount to £2448 per
annum. The rates are one-third of the rent. How much
is the rent per month?

3. Three boys share £2·40. For every 2p A receives,
B receives 3p, and C 7p. How much do they each receive?

4. A, B, and C share a sum of money in the ratio 3:5:7,
respectively. If C receives £4·20, what was the sum?

5. In his will a man left £18 000 to be shared as follows:
one-third to his wife, one-quarter to each of his 2 sons,
and the rest to be divided equally between his 3 daughters.
How much will each daughter receive?

6. A man left his property to be divided as follows:
one-third to his wife, one-quarter of the remainder to each
of his 3 sons, and the rest to be divided equally between
his 2 daughters. If each daughter received £750, what was
the total amount of his property?

7. A, B, and C join in a commercial venture with con-
tributions of £450, £750, and £500 respectively. How
much should they each receive from a net profit of £4250?

8. A and B are in partnership. A is to receive 10% of the profits as manager and the residue is divisible two-thirds to A and one third to B. In what ratio are the profits divisible?

9. A, B, and C hold 25%, 45%, and 30% respectively of the capital of a business. C's investment is £4500. Find the amounts invested by A and B.

10. A commences business with a capital of £2500, 2 months later B enters the business with a capital of £4000, and 4 months after this C enters the partnership with a capital of £2000. The profits are to be divided in proportion to capital. At the end of 12 months the profits were £1435. How should this be divided?

11. A commenced business with a capital of £3500, 2 months later B entered the business with a capital of £3600, and 4 months after this C was taken into the partnership with a capital of £3000. The profits are to be shared equally, but before division of profits each partner is to be credited with 10% per annum interest on capital. The profits at the end of 12 months were £2540. With what amount will each partner's account be credited?

12. The profits of a business are £3500. A is to receive a salary of £1000 and two-fifths of the remaining profits. B is to receive the residue. How much do they each receive?

13. The capitals of 3 partners in a business are: A, £2000; B, £2200; and C, £2300. The only agreement as to the allocation of profit is that A should receive 40% of net profit as manager, and then each partner should receive 5% interest on capital. How should a net profit of £4500 be divided?

14. Three partners made a profit of £1236. A received a half, B one-third, and C the residue. How much did they each receive?

15. A and B are in partnership with capitals of £4000 and £5000 respectively. The net profit is to be divided in the ratio 7:8. The gross profit for the year is £4700. £1400 of this is spent on salaries and £450 on other expenses. How much do they each receive?

CHAPTER 13

Factors and Multiples— Powers and Roots

1. Factors and Multiples

The student should revise the definitions in Chapter 2 before proceeding with this chapter.

A *Factor* or *measure* or *aliquot part* of a number is a number which is contained an exact number of times in that number. Thus, 7 is a factor of 56, so is 2 and 4 and 8.

A *Multiple* is a number which contains another number an exact number of times. Thus 56 is a multiple of 7. It is also a multiple of 2, 4, and 8.

2. Tests for Divisibility

A number is divisible by:

3 or 9, if the sum of its digits is so divisible.
4, if the last 2 digits are so divisible.
8, ,, ,, 3 ,, ,,
6, if it is even and divisible by 3.
11, if the sum of the digits in the odd places equals the sum of the digits in the even places, or if the difference between the sums is divisible by 11 or a multiple of 11.

3. Factorisation

To find the prime factors of a number, it is usually easier to commence with the lowest prime factors.

Ex. 1   Find the prime factors of 13 464.

$$13\,464 = \begin{array}{r|r} 2 & 13464 \\ \hline 2 & 6732 \\ \hline 2 & 3366 \\ \hline 3 & 1683 \\ \hline 3 & 561 \\ \hline 11 & 187 \\ \hline & 17 \end{array}$$

$$= 2 \times 2 \times 2 \times 3 \times 3 \times 11 \times 17$$
$$= 2^3 \times 3^2 \times 11 \times 17.$$

4. H.C.F. and L.C.M.

The *Highest Common Factor* of 2 or more numbers and their *Least Common Multiple* may be obtained in one operation by resolving the numbers into their prime factors.

Ex. 2 Find the H.C.F. and L.C.M. of 504, 936, 792.

$$\begin{array}{r|rrr} 2 & 504, & 936, & 792 \\ \hline 2 & 252, & 468, & 396 \\ \hline 2 & 126, & 234, & 198 \\ \hline 3 & 63, & 117, & 99 \\ \hline 3 & 21, & 39, & 33 \\ \hline & 7, & 13, & 11 \end{array}$$

$$\left. \begin{array}{l} \text{H.C.F.} = 2^3 \times 3^2 \qquad\qquad\qquad\quad = \mathbf{72} \\ \text{L.C.M.} = 2^3 \times 3^2 \times 7 \times 11 \times 13 = \mathbf{72\,072} \end{array} \right\} \textbf{Ans.}$$

The H.C.F. = the product of the *common* factors.
The L.C.M. = the product of *all* the factors.

5. Powers and Roots

The *Power* of a number is the result obtained by multiplying a number by itself a certain number of times. Thus, 2×2 is called the second power or square of 2. $2 \times 2 \times 2$ is called the third power or cube of 2. $2 \times 2 \times 2 \times 2$ is the fourth power of 2.

2 × 2 may be written 2^2.
2 × 2 × 2 may be written 2^3.
2 × 2 × 2 × 2 may be written 2^4.

An *Index* (plural indices) is a small number written to the right and slightly above a number to denote the power to which it is raised. Thus, $3 \times 3 \times 3 \times 3 = 3^4$ and 4 is the power to which 3 is raised. $3^4 = 81$.

A *Root* is the converse of a power. It is the number which, if multiplied by itself a certain number of times, will produce the given number. Thus, 81 is the fourth power of 3, and 3 is the fourth root of 81. 64 is the cube of 4, and 4 is the cube root of 64.

A root is indicated by a radical with a small figure to the left giving the root. Thus, $\sqrt[3]{125}$ indicates the cube root of 125, *i.e.* 5. In square root, the small 2 is usually omitted; thus, $\sqrt{9}$ = the square root of 9, *i.e.* 3.

6. To Find the Square Root by Factors

Express the number in prime factors and divide each index by 2.

Ex. 3   Find the square root of 7056.

2	7056
2	3528
2	1764
2	882
3	441
3	147
7	49
	7

$\sqrt{7056} = \sqrt{2^4 \times 3^2 \times 7^2}$
$= 2^2 \times 3 \times 7$
$= 84$ **Ans.**

Note: Cube root can be found by factorising the number and dividing the indices by 3. This method can only be applied where the number is a perfect square or perfect cube, and it should only be used where the numbers factorise easily.

7. To Find the Square Root, Ordinary Method

Ex. 4   Find the square root of 61 009.

```
                  root
        6,10,09(247
        4
     44│210
        176
    487│ 3409
        3409
```

$$\sqrt{61\ 009} = \mathbf{247} \qquad\qquad \textbf{Ans.}$$

Divide into periods of 2 figures each. In *whole numbers*, count to the *left*.

The first period will contain only 1 number if there is an odd number of digits and 2 numbers if there is an even number of digits in the whole numbers.

In *decimals*, count to the *right*, adding a cipher, if necessary, to complete the last period.

Mental steps:
   (1) Find the nearest root to 6. This is 2. Place 2 in the root. Square 2 (*i.e.* 4) and place under the first period, 6. Subtract. Bring down the next period, *i.e.* 10, making 210.
   (2) On the left, place twice the root already found, *i.e.* 4.
   (3) Find how many times 4 is contained in the dividend *without the last figure*. This is 5. Mentally try 5. This is too large, so try 4.
   (4) Place 4 after the 4 in the first column, making 44.
   (5) Multiply 44 by 4 (last figure in root), making 176.
   (6) Continue as above, each time bringing into the first column twice the root already found and into the partial dividend (the second column) one period of 2 figures.

If a number is not a perfect square, the process may be extended to find a root of any required degree of accuracy by bringing down ciphers, two at a time.

Ex. 5   Find the square root of 1250 to 2 decimal places.

```
12,50(35·355
  9
65⌐ 3 50
    3 25
703⌐ 2500
     2109
7065⌐ 39100
      35325
     ⌐ 3775
```

$$\sqrt{1250} = \mathbf{35 \cdot 36} \qquad \textbf{Ans.}$$

Ex. 6   Extract the square root of 457·836 correct to 3 decimal places.

```
457·8360(21·3971
  4
41⌐ 57
    41
423⌐ 16 83
     12 69
4269⌐ 4 1460
      3 8421
42787⌐ 303900
       299509
        4391
```

$$\sqrt{457 \cdot 836} = \mathbf{21 \cdot 397} \qquad \textbf{Ans.}$$

The following are examples of roots which the student should practise until he can extract them at sight.

1. $\sqrt{4}$	$= 2.$		2. $\sqrt{0 \cdot 04}$	$= 0 \cdot 2.$
3. $\sqrt{0 \cdot 0004}$	$= 0 \cdot 02.$		4. $\sqrt{64}$	$= 8.$
5. $\sqrt{0 \cdot 64}$	$= 0 \cdot 8.$		6. $\sqrt{0 \cdot 0064}$	$= 0 \cdot 08.$
7. $\sqrt{121}$	$= 11.$		8. $\sqrt{0 \cdot 0121}$	$= 0 \cdot 11.$

9. $\sqrt{0\cdot000\,121} = 0\cdot011.$ 10. $\sqrt{900} = 30.$

11. $\sqrt{8100} = 90.$ 12. $\sqrt{12\,100} = 110.$

Ex. 7   Find the greatest sum which is contained exactly in £48·30 and £115·50.

Reduce to 10p's and find the H.C.F.

$$
\begin{array}{r|rr}
3 & 483, & 1155 \\
\hline
7 & 161, & 385 \\
\hline
& 23, & 55
\end{array}
$$

H.C.F. $= 3 \times 7 = 21.$

$\therefore$ Greatest sum $=$ **£2·10**   **Ans.**

Exercise 13

1. Write down the squares of all the numbers from 1 to 20.

Write down the square roots of the following:
2. 121.   3. 256.   4. 2·25.   5. 289.
6. 0·0121.   7. 1·44.   8. 0·0256.   9. 8100.

Resolve the following numbers into prime factors:
10. 216   11. 275.   12. 1728.   13. 1125.   14. 174 240.

Find, by factors, the H.C.F. and L.C.M. of the following.
15. 360 and 504.   16. 792 and 936.   17. 675 and 1575.
18. 432 and 720.   19. 11 088 and 13 104.

Find the H.C.F. of:
20. 567 and 621.   21. 1026 and 1653.
22. 221 and 255.   23. 1903 and 2497.

24. Find the least sum of money which can be divided exactly by: 12p, 21p, 33p, and 42p.

25. What is the greatest sum of money contained exactly in £1·90 and £2·20?

26. Find the greatest weight which is contained an exact number of times in 66 g and 96 g.

Find, to 4 significant figures, the square root of each of the following:

27. 83 754   28. 237 816.   29. 0·0867.   30. 2·83.

Find: (*a*) the prime factors, and (*b*) give the square roots of each of the following:

31. 324.   32. 4356.   33. 1296.
34. 1936.   35. 5929.   36. 1764.

37. A boy saves 12½p a week. In how many weeks will he have saved a sum which is divisible exactly into units of 40p?

38. What is the least number which will give a remainder of 5 when divided by 8, 12, and 27?

CHAPTER 14

Mensuration of Rectangular Surfaces and Solids

A *point* denotes position.

A *straight line* is the shortest distance between two points. A line has *length*, but is said to have *no breadth*. The measurement of lines is called *linear measure*.

A *rectangular surface* has length and breadth, but no thickness.

A *rectangular solid* has length, breadth, and thickness.

When two straight lines meet at a point they are said to form an *angle*.

When two straight lines cross each other so as to form four equal angles, each of the angles is called a *right-angle*.

A *parallelogram* is a figure bounded by four straight lines, whose opposite sides are parallel.

A *rectangle* is a right-angled parallelogram, *i.e.* its four angles are right-angles.

A *rectangular solid* is a body having length, breadth, and thickness, whose opposite faces are equal and parallel rectangles.

The student should carefully consider the following, as the easy solution of the problems which follow depends upon a thorough mastery of this groundwork.

$$\text{linear} \times \text{linear} = \text{square measure or } area.$$

$$\frac{\text{area}}{\text{linear}} = \text{linear}.$$

$$\text{linear} \times \text{linear} \times \text{linear} = \text{cubic measure or } volume.$$

$$\frac{\text{volume}}{\text{linear}} = \text{area}.$$

$$\frac{\text{volume}}{\text{area}} = \text{linear}.$$

The term 'linear' refers to the measurement of a straight line. This may be expressed as length, breadth, thickness, height, or depth according to the nature of the problem.

Ex. 1   How many books, each 2·5 cm thick, can be placed on a shelf which is 2 ft 3 in long? (1 m = 39·37 in.)

$$\text{Number} = \frac{27 \times 100}{39\cdot37 \times 2\cdot5}$$

$$= 27 \qquad \textbf{Ans.}$$

$$
\left[
\begin{array}{l}
27 \\
3937)108000(\\
7874 \\
\overline{29260}
\end{array}
\right]
$$

Mental steps:

1. Divide 27 in by 39·37, which brings it to metres.
2. Multiply by 100 = centimetres.
3. Divide by 2·5 = number of books.

Ex. 2   A rectangular field is 660 m long and 550 m wide. Find its area in hectares.

$$\text{Area} = 660 \times 550 \text{ m}^2$$

$$= \frac{660 \times 550}{10\,000} \text{ ha}$$

$$= \textbf{36·30 ha} \qquad \textbf{Ans.}$$

Ex. 3   A rectangular field has an area of 60 ha. Its length is 1200 m. Find its width.

$$\text{Width} = \frac{60 \times 10\,000}{1200} \text{ m}$$

$$= \textbf{500 m} \qquad \textbf{Ans.}$$

Ex. 4   Find, in cubic metres, the volume of earth which has been removed from a trench, 6 m long, 60 cm wide, and 1·5 m deep.

$$\text{Volume} = 6 \times 0\cdot6 \times 1\cdot5 \text{ m}^3$$

$$= \textbf{5·4 m}^3 \qquad \textbf{Ans.}$$

Ex. 5   The volume of a plank of wood is 27 dm³. It is 3 m long and 30 cm wide. Find the thickness of the plank.

$$\text{Thickness} = \frac{27\,000}{300 \times 30} \text{ cm}$$

$$= \textbf{3 cm} \qquad \textbf{Ans.}$$

Ex. 6 The volume of a stone slab is 750 cm³. Its thickness is 3 cm. Find the area of the slab.

$$\text{Area} = \frac{750}{3} \text{ cm}^2$$
$$= \textbf{250 cm}^2 \qquad \textbf{Ans.}$$

Carpet for floors may be bought in one piece or it may be bought by the length of a certain width and laid in strips. Floor space round the edge of a carpet is usually spoken of as *the surround*. Wallpaper is usually sold in rolls; each roll, or piece, is 11 m in length and 52 cm wide.

Ex. 7 A room is 7·5 m long and 4·5 m wide. A carpet, 5·5 m by 3 m, is placed in the centre. Find the cost of covering the surround with vinyl at 90p a square metre.

$$\begin{aligned}
\text{Area of floor} \quad &= 7\cdot5 \times 4\cdot5 \text{ m}^2 \\
&= 33\cdot75 \text{ m}^2. \\
\text{,,} \qquad \text{carpet} \quad &= 16\cdot5 \text{ m}^2. \\
\text{,,} \qquad \text{surround} \quad &= (33\cdot75 - 16\cdot5) \text{ m}^2 \\
&= 17\cdot25 \text{ m}^2. \\
\text{Cost} &= 17\cdot25 \times 90\text{p} \\
&= \textbf{£15·52}\tfrac{1}{2} \qquad \textbf{Ans.}
\end{aligned}$$

Note: This is mathematical accuracy and assumes (1) that not a square cm of the vinyl will be wasted and (2) that the exact amount could be bought.

In actual practice, we should probably buy an integral number of square metres; the little over would allow for wastage.

We should probably buy 18 m², and the cost would then be:

$$90\text{p} \times 18 = £16\cdot20.$$

The area of the walls of a rectangular room is:

Perimeter of room × height,
or (length + breadth) × 2 × height.

The *perimeter* is the boundary line, or the line enclosing an area. In the case of a rectangular area it is (length + breadth) × 2.

Ex. 8   A room is to be papered with a plain paper, so there will be very little wastage. In pattern paper there is a much greater wastage in matching up the pattern. The dimensions of the room are: length, 7 m, width, 4·5 m, height for papering, 2·5 m. There is a French window of area 4 m²; a door of area 2 m²; and a fireplace of area 2 m². What will be the cost of papering the walls with wallpaper which is sold by the piece of 11 m in length and is 52 cm wide and costs £4·50 a piece?

Area of walls = $(7 + 4·5) \times 2 \times 2·5$ m²
= 57·5 m².

French window	4 m².
Door	2 ,,
Fireplace	2 ,,

8 m².

Area to be covered   = 49·5 m².

Length of paper required = $\dfrac{49·5}{0·52}$ m.

Number of pieces   = $\dfrac{49·5}{0·52 \times 11} = 9.$

Cost = £4·50 × 9 = **£40·50**   **Ans.**

Ex. 9   On the floor of a room there is a carpet, 3 m by 4 m, and there is an uncarpeted surround 50 cm wide. Find the area of the surround.

Area of floor	= 5 × 4 m²	
	= 20 m².	
Area of carpet	= 12 ,,	
Area of surround	= **8 m²**	**Ans.**

Note: It is often a help in problems of this type to draw a rough diagram. This should not necessarily be to scale, as its function is that of a visual aid in placing the dimensions.

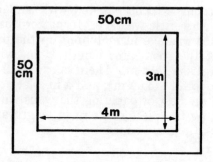

Ex. 10 A hall is 13 m by 17 m. Find the cost of carpeting the floor (leaving a surround, 2 m wide) with carpet, 2·25 m wide, which costs £10·50 a metre.

$$\text{Number of strips in 9 m, 4.}$$
$$\text{Length of strip, 13 m.}$$
$$\text{Length required} = 13 \times 4 \text{ m}$$
$$= 52 \text{ m.}$$
$$\text{Cost} = £10·5 \times 52$$
$$= £546 \qquad \textbf{Ans.}$$

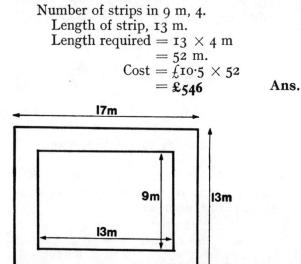

Ex. 11 Find the weight, in kilogrammes, of a metre length of an iron girder which has an L-shaped cross-section. The inner length of the vertical arm is 6 cm, the inner length of the horizontal arm is 4·5 cm, and the width of each arm of the L is 1·5 cm. The weight of 1 cm³ of the iron being 8·5 grammes.

Area of cross section
$$= (6 + 6) \times 1 \cdot 5 \text{ cm}^2$$
$$= 18 \text{ cm}^2.$$
Volume $= 100 \times 18 \text{ cm}^3.$
Weight $= 1800 \times 8 \cdot 5 \text{ g}$
$$= \frac{1800 \times 8 \cdot 5}{1000} \text{ kg}$$
$$= \textbf{15} \cdot \textbf{3 kg} \qquad \textbf{Ans.}$$

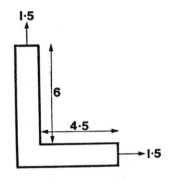

Ex. 12   An iron rod of square section of side 3 cm weighs 75 kg. Given that 1 cm³ of iron weighs 8 g, find the length of the rod.

Volume $= \dfrac{75\,000}{8} \text{ cm}^3.$

(*Note:* Volume varies in direct proportion to weight.)

Section area $= 3 \times 3 \text{ cm}^2$
$$= 9 \text{ cm}^2.$$

Length $= \dfrac{\overset{6250}{\cancel{75\,000}}}{\underset{2}{8} \times \underset{3}{9}} \text{ cm}$

$$= 1042 \text{ cm}$$
$$= \textbf{10} \cdot \textbf{42 m} \qquad \textbf{Ans.}$$

Ex. 13   A sheet of lead, 3 mm thick, weighs 15 kg. If 1 cm³ of lead weighs 10 g, find the area of the sheet.

$$\text{Volume} = \frac{15\ 000}{10}\ \text{cm}^3.$$

$$\text{Area} \quad = \frac{1500}{0\cdot3}\ \text{cm}^2$$
$$= \mathbf{5000\ cm^2} \qquad \textbf{Ans.}$$

Ex. 14   A rectangular slab of marble, 3 m long, 1 m wide, and 2 cm thick, weighs 162 kg. Find the weight of 1 cubic metre of the marble.

$$\text{Volume} \qquad = 3 \times 0\cdot02\ \text{m}^3$$
$$= 0\cdot06\ \text{m}^3.$$

$$\text{Weight of 1 m}^3 = \frac{16\ 200}{6}\ \text{kg}$$
$$= 2700\ \text{kg}$$
$$= \mathbf{2\cdot7\ t} \qquad \textbf{Ans.}$$

Note: To find the weight of a unit, given a volume and weight of a volume, divide the weight by the volume expressed in terms of the unit.

Thus, if $\frac{2}{3}$ dm³ weighs 180 kg

then 1 dm³ weighs $\frac{180}{1} \div \frac{2}{3}$ kg
$$= \frac{180}{1} \times \frac{3}{2}\ \text{kg} = \mathbf{270\ kg}$$

If $1\frac{1}{2}$ dm³ weighs 180 kg

then 1 dm³ weighs $\frac{180}{1} \div \frac{3}{2}$ kg
$$= \frac{180}{1} \times \frac{2}{3}\ \text{kg} = \mathbf{120\ kg}$$

Exercise 14

1. How many books, $\frac{3}{4}$ in thick, can be placed on a shelf, 1 m long? (1 m = 39·37 in.)

2. A field is 605 m long and 440 m wide. Find its area in hectares.

3. A square field has an area of 5 ha. Find the length of a side. (Answer to the nearest metre.)

4. A rectangular field has an area of 35 ha. Its width is 308 m. Find its length.

5. A plank of wood is 3 m long, 9 cm wide, and 2 cm thick. Find its volume in cubic decimetres.

6. 4·5 dm³ of iron is cast into an iron rail which has a cross section of $2\frac{1}{4}$ cm². Find the length of the rail in metres.

7. A marble slab has a volume of 3 dm³. Its thickness is 2 cm. Find the area of the slab.

8. A room is 7 m long and 5 m wide. A carpet, 3 m by 4 m, is placed in the centre. Find the cost of staining the surround at 7p a square metre.

9. A man decides to paper a room, 5 m long, 4 m wide, and 2·5 m high. Of this area, the door takes up 2 m²; the fireplace, 2·5 m²; and 2 windows, each 1·5 m². The wallpaper is sold in rolls of 11 m in length and is 52 cm wide. It is £5·10 a roll. Find the cost of the wallpaper.

10. On the floor of a room a carpet, 3·5 m by 3 m, is placed. There is a surround, 50 cm wide. Find the cost of covering the surround with vinyl at £2·50 a square metre. (Purchase to nearest square metre above.)

11. A hall, 60 m by 16 m, is to be covered with carpet which is 50 cm wide and costs £4·05 a metre. Find the cost of the carpet.

12. Find the weight, in kilogrammes, of 10 m length of a steel girder which has a T-shaped cross-section. The inner length of the vertical arm of the T is 7 cm, and the length of the horizontal top of the T is 5·5 cm. The width throughout is 1·5 cm. The weight of 1 cm³ of the steel is 9·8 grammes.

13. An iron rod of square cross-section of side 2 cm weighs 72 kg. Given that the iron weighs 8 g/cm³, find the length of the rod in metres.

14. A sheet of lead, 1·5 m long and 48 cm wide, weighs 36 kg. Given that lead weighs 12 g/cm³, find, to the nearest mm, the thickness of the sheet.

15. A plank of wood, 4 m long, 20 cm wide, and 5 cm thick, weighs 32 kg. Find the weight of 1 cubic decimetre of the wood.

16. Find the cost of laying a kitchen floor, 4 m by 2·5 m, with square tiles of 20 cm side, which cost £1·80 for ten.

17. Find the cost of paving a passage, 9 m long and 2 m wide, with stones measuring 12 cm by 20 cm, and costing £2·30 for ten.

18. A rectangular courtyard is 29·75 m long and 12·25 m wide. Find: (*a*) the length of the side of the largest square paving stone that can be used to pave the courtyard, and (*b*) the number of stones required.

19. Find the cost of turfing a lawn, 8 m by 15 m. Each turf is 40 cm by 1 m, and the price is £14 per 100.

20. Find the cost of excavating a trench, 80 m long, 1·4 m deep, and 75 cm wide, at 70p per cubic metre.

21. Find the weight in tonnes of an iron girder, 120 m long, with cross-section 56 cm², given that the iron weighs 7·5 g/cm³.

CHAPTER 15

Mensuration of the Triangle and Triangular Prism

1. Triangle

A *Triangle* is a figure bounded by 3 straight lines.

To find the area of a triangle.

The area of a triangle is half the area of a rectangle having length and breadth corresponding to base and height of the triangle (*see* Fig. 1).

(1) *Given base and perpendicular height.*

Area = $\frac{1}{2}$ (base × height).

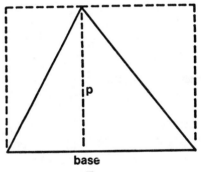

FIG. 1.

Ex. 1   Find the area of a triangle which has a base of 8 cm and a perpendicular height of 5·5 cm.

$$\text{Area} = \frac{1}{2}(8 \times 5\cdot5) \text{ cm}^2$$
$$= \textbf{22 cm}^2 \qquad \textbf{Ans.}$$

(2) *Given length of the 3 sides.*

$$\text{Area} = \sqrt{s(s-a)(s-b)(s-c)}$$

where $s = \frac{1}{2}$ sum of sides a, b, c.

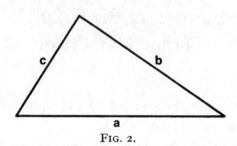

FIG. 2.

Ex. 2   Find, to 1 decimal place, the area of a triangle with sides 4 cm, 7 cm, and 9 cm.

(Sum of sides = 4 + 7 + 9 = 20 cm.)

$\text{Area} = \sqrt{10(10-4)(10-7)(10-9)} \text{ cm}^2$

$= \sqrt{10 \times 6 \times 3} \text{ cm}^2$

$= \sqrt{180} \text{ cm}^2$

$= \mathbf{13 \cdot 4} \text{ cm}^2$ **Ans.**

$$\begin{array}{r}
)180(13\cdot4 \\
1 \\
23\overline{)\,80} \\
69 \\
264\overline{)1100} \\
1056 \\
\hline
44
\end{array}$$

2. Triangular Prism

A *Prism* may be defined as a solid with end faces exactly alike and whose sides are parallelograms. If the edges are perpendicular to the end faces, the prism is called a *right* prism, otherwise it is called oblique. The word 'prism' used without qualification usually means a right prism. The shape of the end, or cross-section, gives the name to the prism. Rectangular prisms were discussed in the previous chapter under the name of rectangular solids. Fig. 3 represents a triangular prism.

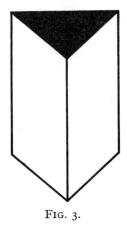

Fig. 3.

In any right prism:

Volume = Area of end (or cross-section) × length,
        or Area of base × height.
Surface area = Area of lateral surfaces + area of 2 ends.
        = Perimeter of end × length + area of 2
           ends.

This is true of any shape of cross-section, provided that it
is uniform.

Ex. 3 Find: (*a*) the volume, and (*b*) the surface area
of a right triangular prism of length 8 cm. Cross-section
dimensions are 2·5 cm, 3·5 cm, and 4 cm. (Answer in
each case to 3 significant digits.)

Section area $= \sqrt{5 \times 2 \cdot 5 \times 1 \cdot 5}$ cm²
            $= 4 \cdot 33$ cm².
Volume $= 4 \cdot 33 \times 8$ cm³
            $= 34 \cdot 64$ cm³
            $= \mathbf{34 \cdot 6}$ **cm³**   **Ans.** (*a*)

$$\begin{array}{r} 18 \cdot 75(4 \cdot 33 \\ 16 \\ 83 \overline{\smash{\big)}\ 2\ 75} \\ 2\ 49 \\ 863 \overline{\smash{\big)}\ 26\ 00} \\ 25\ 89 \end{array}$$

Surface area $= 2(4 \cdot 33) + 8(2 \cdot 5 + 3 \cdot 5 + 4)$
            $= \mathbf{88 \cdot 7}$ **cm²**   **Ans.** (*b*)

3. Right-angled Triangle

A *Right-angled Triangle* is one which contains one right-angle.

The *hypotenuse* is the longest side.

The *base* and *perpendicular* are the two shorter sides.

The square of the hypotenuse is equal to the sum of the squares of the base and perpendicular (*see* Fig. 4).

Given any two sides of a right-angled triangle, the third can be found.

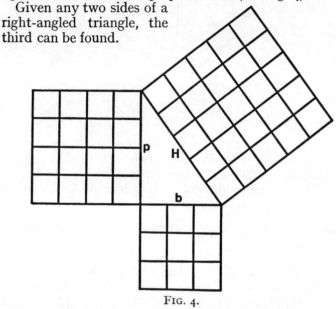

FIG. 4.

Let $p =$ perpendicular.
        $b =$ base.
        $H =$ hypotenuse.

Then, $H^2 = p^2 + b^2 \quad = 5^2 = 4^2 + 3^2$
$$\qquad\qquad\quad = 25 = 16 + 9.$$
$$H = \sqrt{p^2 + b^2} = \sqrt{25} \qquad = 5.$$
$$p = \sqrt{H^2 - b^2} = \sqrt{25 - 9} = 4.$$
$$b = \sqrt{H^2 - p^2} = \sqrt{25 - 16} = 3.$$

If two equal right-angled triangles be joined (*see* Fig. 5) we have a rectangle with length and breadth corresponding to the perpendicular and base of the triangles, while the diagonal corresponds to the hypotenuse of the triangles.

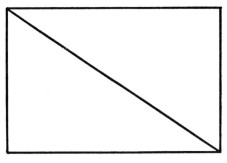

FIG. 5.

In finding base and perpendicular it should be remembered that the difference between the squares of two numbers is equal to the product of their sum and difference.

$$A^2 - B^2 = (A + B)(A - B).$$

Thus,
$$34^2 - 32^2 = (34 + 32) \times (34 - 32)$$
$$= 66 \times 2$$
$$= \mathbf{132}$$

Ex. 4 The base and perpendicular of a right-angled triangle are 4·5 cm and 6 cm respectively. Find the length of the hypotenuse.

$$\text{Hypotenuse} = \sqrt{4 \cdot 5^2 + 6^2} \text{ cm}$$
$$= \sqrt{56 \cdot 25} \text{ cm}$$
$$= \mathbf{7 \cdot 5 \text{ cm}} \qquad \textbf{Ans.}$$

Ex. 5 The hypotenuse of a right-angled triangle is 13 cm and the base is 5 cm. Find the length of the perpendicular.

$$\text{Perpendicular} = \sqrt{13^2 - 5^2} \text{ cm}$$
$$= \sqrt{(13 + 5)(13 - 5)} \text{ cm}$$
$$= \sqrt{18 \times 8} \text{ cm}$$
$$= \sqrt{144} \text{ cm}$$
$$= \textbf{12 cm} \qquad \textbf{Ans.}$$

Ex. 6   A rectangle is 7·5 cm long and 4 cm wide. Find the length of its diagonal.

$$\text{Length} = \sqrt{7\cdot5^2 + 4^2} \text{ cm}$$
$$= \textbf{8·5 cm} \qquad \textbf{Ans.}$$

Ex. 7   The length of a straight path, running diagonally across a rectangular field, is 120 m. The length of one side of the field is 95 m. Find, to the nearest m, the length of the other side.

$$\text{Length} = \sqrt{120^2 - 95^2} \text{ m}$$
$$= \sqrt{(120 + 95)(120 - 95)} \text{ m}$$
$$= \sqrt{215 \times 25} \text{ m}$$
$$= \textbf{73 m} \qquad \textbf{Ans.}$$

Ex. 8   A man places a 3·9 m ladder against a wall. If he places it 1·5 m from the base, at what height will it touch the wall?

$$\text{Height} = \sqrt{3\cdot9^2 - 1\cdot5^2} \text{ m}$$
$$= \sqrt{5\cdot4 \times 2\cdot4} \text{ m}$$
$$= \sqrt{12\cdot96} \text{ m}$$
$$= \textbf{3·6 m} \qquad \textbf{Ans.}$$

Ex. 9   A rope is stretched from the top of a ship's mast to a point on deck 13·5 m from the base of the mast. The length of the rope is 22·5 m. What is the height of the mast?

$$\text{Height} = \sqrt{22 \cdot 5^2 - 13 \cdot 5^2} \text{ m}$$
$$= \sqrt{36 \times 9} \text{ m}$$
$$= \textbf{18 m} \qquad \textbf{Ans.}$$

$$\left[\begin{array}{l})324(18 \\ \quad 1 \\ 28\overline{\smash{)}224} \\ \quad 224 \end{array}\right]$$

Note: $\sqrt{36 \times 9} = \sqrt{6^2 \times 3^2} = 6 \times 3 = \textbf{18}$.

Exercise 15

1. Find the area of a triangle of 17·5 cm base and a perpendicular of 8 cm.

2. A field is in the shape of a right-angled triangle, with base 56 m and perpendicular 42 m. Find the length of the third side.

3. Find, by two methods, the area of a triangular field with sides of 42 m, 56 m, and 70 m.

4. A triangle has sides of 5 cm, 6 cm, and 7 cm. Find, to a tenth of a cm², the area of the triangle.

5. A right-angled triangle has a base of 12 cm and a height of 9 cm. Find the length of the hypotenuse.

6. The base and hypotenuse of a right-angled triangle are 8 cm and 10 cm respectively. Find the height of the triangle.

7. A rectangle is 15 cm by 8 cm. Find the length of its diagonal.

8. Find the volume of a right prism 8 cm long, the area of its cross-section being 2·5 cm².

9. A straight path running diagonally across a rectangular field is 100 m long. The length of one side of the field is 75 m. Find, to the nearest metre, the length of the other side.

10. The area of a square field is 10 ha. Find, to the nearest metre, the length of a straight path running diagonally across the field.

11. A bedroom window-ledge is 4 m from the ground. At what distance from the base should a 4·5 m ladder be placed so that it will reach the window-ledge? (Answer to the nearest decimetre.)

12. Find the area of a square garden which has a diagonal of 20 m.

13. Find the weight in kilogrammes of an iron girder, a cross-section of which is in the form of a triangle of sides 4 cm, 5 cm, and 6 cm in length, given that 1 cm³ of the iron weighs 7·5 g and the length of the girder is 5 m.

14. A man wishes to make a ladder which will reach the top of a wall, 6 m high, when placed 2 m from the base of the wall. Find, to the nearest decimetre above, the length of the ladder.

15. Find the surface area of a right triangular prism, the cross-section dimensions being 1·5 cm, 2·5 cm, and 3 cm, and the length, 6 cm. (Answer to 3 significant digits.)

16. Find the area of the field in Question 1, Exercise 11. (Answer to the nearest square metre.)

CHAPTER 16

Mensuration of the Circle and Cylinder

1. Circle

The *Circumference* of a circle is the distance round it, or the boundary line.

The *Diameter* is the distance from any point on the circumference, through the centre, to another point on the opposite side of the circumference.

The *Radius* is the shortest distance from the centre to any point on the circumference.

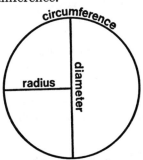

The circumference of a circle bears to its diameter a *constant ratio* which is denoted by the symbol π (pi), and is approximately $3\frac{1}{7}$ or $3 \cdot 1416$.

Unless stated to the contrary, π will be taken as $3\frac{1}{7}$ in the following exercises.

$$\text{Circumference} = \text{diameter} \times \pi$$
$$= \pi D$$
$$= 2\pi r.$$
$$\text{Area} = (\text{radius})^2 \times \pi$$
$$= \pi r^2.$$

Ex. 1 The diameter of a circle is 7 cm. Find its circumference.

$$\text{Circumference} = \tfrac{7}{1} \times \tfrac{22}{7} \text{ cm}$$
$$= \textbf{22 cm} \qquad \textbf{Ans.}$$

Ex. 2   The circumference of a circle is 16·5 cm. Find its diameter.

$$\text{Diameter} = \tfrac{33}{2} \div \tfrac{22}{7} \text{ cm}$$
$$= \tfrac{33}{2} \times \tfrac{7}{22} \text{ cm}$$
$$= \textbf{5·25 cm} \qquad \textbf{Ans.}$$

Ex. 3   The diameter of a circle is 1 m. Find its area as the decimal of a square metre. ($\pi = 3\cdot 1416$.)

$$\text{Area} = \tfrac{1}{2}^2 \times 3\cdot1416 \text{ m}^2$$
$$= \tfrac{1}{2} \times \tfrac{1}{2} \times 3\cdot1416 \text{ m}^2$$
$$= \textbf{0·7854 m}^2 \qquad \textbf{Ans.}$$

Note: From the above it will be seen that the area of a circle is:

$$\frac{3\cdot1416}{4} \text{ times the diameter squared} = \frac{\pi D^2}{4}$$
$$= D^2 \times 0\cdot7854.$$

Ex. 4   The area of a circle is 11 cm². Find, to 1 decimal place, the length of its diameter in centimetres.

$$\pi r^2 = \text{area}.$$
$$r^2 = \frac{\text{area}}{\pi}. \quad \text{(Dividing both sides by } \pi.)$$
$$r = \sqrt{\frac{\text{area}}{\pi}}. \quad \text{(Sq. root of both sides.)}$$
$$\therefore \text{Radius} = \sqrt{\frac{11}{\pi}} \text{ cm}$$
$$= \sqrt{\frac{11 \times 7}{22}} \text{ cm}$$
$$= 1\cdot87 \text{ cm.}$$
$$\text{Diameter} = 1\cdot87 \times 2 \text{ cm}$$
$$= \textbf{3·7 cm} \qquad \textbf{Ans.}$$

$$\left[\begin{array}{l} \quad)3\cdot5(1\cdot87 \\ \quad\quad 1 \\ 28 \overline{\smash)2\ 50} \\ \quad\quad 2\ 24 \\ 367 \overline{\smash)2600} \\ \quad\quad 2569 \end{array}\right]$$

Or: $\pi = 3\cdot1416$.

$$0\cdot7854D^2 = \text{area}$$

$$D^2 = \frac{\text{area}}{0\cdot7854}$$

$$D = \sqrt{\frac{\text{area}}{0\cdot7854}}$$

$$= \sqrt{\frac{11}{0\cdot7854}}$$

$$= \mathbf{3\cdot7 \ cm} \qquad \textbf{Ans.}$$

$$\begin{array}{r}
14 \\
714)\overline{10000(} \\
714 \\
\hline
2860 \\
2856 \\
\hline
4
\end{array}$$

$$\begin{array}{r}
)14(3\cdot74 \\
9 \\
67 \ \overline{\big)500} \\
469 \\
744 \ \overline{\big)3100} \\
2976
\end{array}$$

Ex. 5   The diameter of the wheels of a car is 56 cm. Find the revolutions per minute made by a wheel when the car is travelling at 33 km/h.

$$\begin{aligned}
\text{Circumference of wheel} \quad &= 56\pi \text{ cm} \\
&= 176 \text{ cm.} \\
\text{Distance travelled in 1 min} &= \frac{33 \times 100\,000}{60} \text{ cm.} \\
\text{Number of revs. per min} \quad &= \frac{33 \times 100\,000}{60 \times 176} \\
&= \frac{625}{2} \\
&= 312\tfrac{1}{2} \\
&= \mathbf{313 \ approx.} \qquad \textbf{Ans.}
\end{aligned}$$

2. Annulus or Ring

Concentric circles are circles having a common centre.

An *Annulus* is a figure bounded by two concentric circles. It is commonly referred to as a ring.

The area of a ring = area of larger circle *minus* area of smaller circle.

Let R = radius of larger circle.

Let r = radius of smaller circle.

$$\text{Area} = \pi R^2 - \pi r^2$$
$$= \pi (R^2 - r^2)$$
$$= \pi (R + r)(R - r).$$

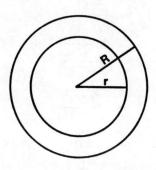

Ex. 6 Find the area of a ring, 1·5 cm wide, which has an inner radius of 2·75 cm.

Outer radius = $4\frac{1}{4}$ cm.
Area = $\frac{22}{7}(4\frac{1}{4} + 2\frac{3}{4})(4\frac{1}{4} - 2\frac{3}{4})$ cm²
$= \frac{22}{7} \times \frac{7}{1} \times \frac{3}{2}$ cm²
= **33 cm²** **Ans.**

Ex. 7 The inside and outside circumferences of a circular race track are 572 m and 616 m respectively. Find the width of the track.

Outside diameter = $\dfrac{616 \times 7}{22}$ m
= 196 m.
Inner diameter = $\dfrac{572 \times 7}{22}$ m
= 182 m
Difference = 14 m
∴ Width of track = **7 m** **Ans.**

Or:

$$\text{Difference in diameters} = \frac{616}{\pi} - \frac{572}{\pi} \text{ m}$$

$$= \frac{44}{\pi} \text{ m}$$

$$= 14 \text{ m}.$$

$$\therefore \text{ Width of track} = \textbf{7 m} \qquad \textbf{Ans.}$$

To find the area of a solid cylinder:

Area of curved surface = circumference × height
$$= 2\pi rh, \text{ when } h = \text{height.}$$
Total area (including 2 ends) $= 2\pi rh + 2\pi r^2$
$$= 2\pi r(h + r).$$

To find the volume of a solid cylinder:

Volume = area of base × height
$$= \pi r^2 h.$$

Volume of cylindrical shell (or tube or pipe)

= volume of solid cylinder *minus* volume
of hollow centre.
$$= \pi R^2 h - \pi r^2 h$$
$$= \pi h(R^2 - r^2)$$
$$= \pi h(R + r)(R - r).$$

Ex. 8   Find the total surface area of a cylinder, 17 cm high and 8 cm diameter.

$$\text{Surface area} = 2\pi r(17 + 4) \text{ cm}^2$$

$$= \frac{2 \times 22 \times 4 \times 21}{7} \text{ cm}^2$$

$$= 22 \times 24 \text{ cm}^2$$

$$= \textbf{528 cm}^2 \qquad \textbf{Ans.}$$

Ex. 9   Find the weight of a solid iron rod, 12 m long and with cross-section of 35 mm diameter, given that the iron weighs 8 g/cm³.

$$\text{Volume} = \frac{7}{4} \times \frac{7}{4} \times \frac{22}{7} \times \frac{\overset{75}{\cancel{1200}}}{1} \text{ cm}^3$$

$$\text{Weight} = \frac{7 \times 22 \times \overset{3}{\cancel{75}} \times \overset{2}{\cancel{8}}}{\cancel{1000}} \text{ kg}$$

$$= \textbf{92·4 kg} \qquad\qquad \textbf{Ans.}$$

Ex. 10 A cylindrical tank, 35 cm in diameter, contains 100 litres of water. Find, to the nearest cm, the depth of the water. (1 l = 1 dm³.)

Volume of water = 100 dm³.

$$\text{Area of base} = \frac{7}{4} \times \frac{7}{4} \times \frac{22}{7} \text{ dm}^2.$$

$$\text{Depth} = \frac{100}{1} \div \frac{77}{8} \text{ dm}$$

$$= \frac{100 \times 8}{77} \text{ dm}$$

$$= \textbf{1·04 m} \qquad\qquad \textbf{Ans.}$$

Ex. 11 Find the cost of 100 m of copper wire, 2 mm in diameter, given that copper weighs 12 g/cm³, and the wire costs £3 a kg.

$$\text{Volume} = \frac{22}{7} \times \frac{100\ 000}{1000} \text{ cm}^3.$$

$$\text{Cost} = £\frac{22 \times 100 \times 12 \times 3}{7 \times 1000}$$

$$= \textbf{£11·31} \qquad\qquad \textbf{Ans.}$$

Ex. 12 A bath water-pipe is 1 cm in internal diameter and water flows through at the rate of 8 km/h. If the tap is left on for 6 min, how many litres of water will the bath contain?

(6 min = $\frac{1}{10}$ h, ∴ the volume is equal to a solid cylinder, $\frac{1}{10}$ of 8 km long.)

$$\text{Volume} = \frac{\text{I}}{400} \times \frac{22}{7} \times \frac{\overset{20}{\cancel{80\,000}}}{\cancel{10}} \text{ dm}^3$$

$$= \frac{440}{7} \text{ dm}^3$$

$$= \textbf{63 l} \qquad\qquad \textbf{Ans.}$$

Exercise 16

1. The diameter of a circle is 10·5 cm. Find its circumference.

2. The circumference of a circle is 77 cm. Find its diameter.

3. What will be the cost of putting a fence round a circular pond of diameter 35 m, if the fencing costs £3 a metre?

4. The spokes of a wheel extend to within 2 cm of the outer rim and 2 cm from the axis. Find the circumference of the wheel if the spokes are 24 cm long.

5. The diameter of a circle is 5 cm. Find its area, to the nearest square centimetre.

6. Find the area, to the nearest square decimetre, of a circular flower bed, which has a diameter of 3·5 m.

7. A pond has a surface area of 1936 m². Find the diameter of the pond.

8. The diameter of the wheels of a car is 63 cm. Find the revolutions per minute of a wheel when the car is travelling at 45 km/h.

9. Find the area of a ring, 3 cm wide, which has an outer diameter of 17 cm.

10. An iron washer is a flat, perforated cylinder of thickness 2 mm, with external diameter of 2 cm and internal diameter of 8 mm. Given that 1 cm³ of iron weighs 8 g, find, to the nearest gramme, the weight of 100 of these washers.

11. A circular pond, 48 m in diameter, has a path, 1 m wide, all round. What will be the cost of laying the path with gravel at £2 a square metre?

12. What area of sheet tin will be required to make an oil drum, 2 m high and 1·6 m in diameter?

13. Find the weight of 35 m of lead pipe of external diameter 18 mm and internal diameter 12 mm, given that lead weighs 12 g/cm³.

14. A cistern contains 80 l of water. The outlet pipe is 1 cm in internal diameter. How long will it take to empty the cistern if the water flows through at the rate of 2 m/s?

15. Find, in kilogrammes, the weight of a kilometre length of copper wire of circular section, 4 mm in diameter, given that 1 cm³ of copper weighs 10 g.

16. A cylindrical tank, 1 m in diameter, contains 75 l of water. Find, to the nearest centimetre, the depth of the water.

17. The wheels of a child's bicycle turn 840 times in going 1 km. Find the height of the wheel hubs from the ground.

CHAPTER 17

Foreign Currencies—Exchange

Exchange is the term used for the conversion of sums of money from the coinage of one country to that of another.

The Course of Exchange is the rate ruling at any time. Under normal conditions the variation is slight, but the variations are considerable when conditions are abnormal. Since the Second World War the exchanges have been in a state of flux.

The following gives the unit and sub-unit of coinage of the countries named:

Country					
United Kingdom .	.	.	.	Pound	= 100 pence.
Australia .	.	.	.	Dollar	= 100 cents.
New Zealand	.	.	.	,,	= ,,
Canada .	.	.	.	,,	= ,,
U.S.A. .	.	.	.	,,	= ,,
Hong Kong .	.	.	.	,,	= ,,
Malaysia .	.	.	.	,,	= ,,
South Africa	.	.	.	Rand	= ,,
Sierra Leone	.	.	.	Leone	= ,,
Ghana .	.	.	.	Cedi	= 100 pesewas.
India .	.	.	.	Rupee	= 100 naya paise.
France .	.	.	.	Franc	= 100 centimes.
Belgium * .	.	.	.	,,	= ,,
Switzerland .	.	.	.	,,	= ,,
Germany .	.	.	.	Mark	= 100 pfennige.
Italy .	.	.	.	Lira	= 100 centesimi.
Holland .	.	.	.	Florin	= 100 cents.
Denmark .	.	.	.	Krone	= 100 öre.
Norway .	.	.	.	Krone	= ,,
Sweden .	.	.	.	Krona	= ,,
Spain .	.	.	.	Peseta	= 100 centimos.
Portugal .	.	.	.	Escudo	= 100 centavos.

* 1 belga = 5 francs.

It will be observed that the sub-unit is one hundredth part of the unit. This simplifies considerably all money operations.

The method of quoting the rate of exchange in the case of India is to state the value of the rupee in pence. Thus, we should speak of the rupee as being 7p. In the other cases of the above-named countries, we take the number of foreign coins in relation to a unit of £1. Thus, the Paris rate might be 11·88, *i.e.* £1 = 11·88 francs.

Where the method of quotation is in units of foreign currency to the £1:

To convert £'s to foreign currency:
  Multiply £'s by the rate of exchange.

To convert foreign currency to £'s:
  Divide foreign currency by the rate of exchange.

Ex. 1 Convert 1000 pesetas to £'s when £1 = 184 pesetas.

$$1000 \text{ pesetas} = £\frac{\overset{125}{\cancel{1000}}}{\underset{23}{\cancel{184}}}$$

$$= £5\cdot43 \quad \textbf{Ans.}$$

$$\begin{array}{r} 23)125(5\cdot43 \\ \underline{115} \\ 100 \\ \underline{92} \\ 80 \\ 69 \end{array}$$

Ex. 2 Convert £4·90 to Belgian francs, when £1 = 74 francs.

$$£4\cdot90 = 4\cdot9 \times 74$$
$$= \textbf{362·6 francs} \qquad \textbf{Ans.}$$

Ex. 3 Convert £150 to dollars, given £1 = \$1·85.
$$£150 = 150 \times 1\cdot85$$
$$= 1\tfrac{1}{2} \times 185$$
$$= \textbf{\$277·50} \qquad \textbf{Ans.}$$

Ex. 4 Convert 1000 dollars to sterling when the rate of exchange is 1·875.

$$\$1000 = \pounds\frac{1000}{1\cdot875}$$

$$= \pounds\frac{1600}{3}$$

$$= \pounds533\cdot33 \qquad \textbf{Ans.}$$

Ex. 5 Convert 5000 rupees to sterling when the rupee is worth $7\frac{1}{2}$p.

$$5000 \text{ rupees} = 7\frac{1}{2}\text{p} \times 5000$$
$$= \pounds375 \qquad \textbf{Ans.}$$

Ex. 6 A merchant in India arranges for his wife in England to be credited with 7200 rupees per annum, payable monthly. By how much would her monthly income be increased by a rise in the value of the rupee from 7p to $7\frac{1}{2}$p?

$$\text{Increase for each rupee} = \tfrac{1}{2}\text{p}$$
$$\text{Monthly increase} = \pounds\frac{7200 \times \frac{1}{2}}{12 \times 100}$$
$$= \pounds3 \qquad \textbf{Ans.}$$

Ex. 7 A man spent a holiday in Spain, taking with him £120, which he changed to pesetas at 184 to the £1. He spent 19 000 pesetas and changed the remainder to £'s at 180 to the £1. How much did the holiday cost him?

$$£120 \text{ at } 184 \text{ pesetas} = 22\ 080 \text{ pesetas}$$
$$\text{Amount spent} = 19\ 000 \quad ,,$$
$$\text{Remainder} = 3080 \quad ,,$$

$$3080 \text{ pesetas at } 180 = \frac{\overset{154}{\cancel{3080}}}{\underset{9}{\cancel{180}}}$$

$$= £17\cdot11$$

$$\left[\begin{array}{l} 9)154(17\cdot11 \\ \underline{9} \\ 64 \\ \underline{63} \\ 10 \\ 9 \end{array}\right]$$

Cost of holiday $= £120 - £17\cdot11 = \pounds102\cdot89$ **Ans.**

Ex. 8   An Englishman sold a car in Spain for 450 000 pesetas. When he was paid the exchange rate had risen from 180 to 183. How much did he lose?

$$\text{At first rate } £\frac{450\ 000}{180} = £2500$$

$$\text{At second rate } £\frac{450\ 000}{183} = £2459 \text{ (to nearest £1)}$$

$$\text{Loss} = \textbf{£41} \qquad\qquad \textbf{Ans.}$$

Ex. 9   When the rate of exchange fell from 184 to 180, how much extra had to be paid in sterling for a Spanish shawl, priced at 2760 pesetas?

$$\text{At } 184: £\frac{2760}{184} = £15$$

$$\text{At } 180: £\frac{2760}{180} = £15\cdot33$$

$$\text{Amount extra} = \textbf{33p} \qquad\qquad \textbf{Ans.}$$

Exercise 17

1. Convert 50 francs to £'s, when £1 = 11·00 francs.

2. Convert £1050 to francs, when £1 = 10·85 francs.

3. Convert £250 to dollars, when the rate of exchange is 1·82.

4. Convert 3000 dollars to £'s, given £1 = 1·8750 dollars.

5. Convert £1200 to rupees, given 1 rupee is worth 7½p.

6. Convert 4000 rupees to £'s, when the rupee is worth 8p.

7. A man spent a holiday in France and took with him £100, which he changed to francs at 10·8 to the £. He spent 900 francs and changed the remainder to English money at 11·00 francs to the £. How much did the holiday cost him?

8. A debt of £100 is paid in francs valued at 9p each when francs were 10·8 to the £. How much did the debtor gain or lose?

9. A merchant in England received 20 000 francs when the rate was 8·54. He changed it when the rate was 8·50. How much did he gain by waiting?

10. An Englishman receives an Indian pension of 800 rupees a month. How much will he gain per annum if the rupee rises from 7p to 7½p?

11. An Englishman sold a car in Spain for £1500 in June, when the value of the £ was 180 pesetas. He was paid in August, when the value of the £ was 184 pesetas. Find, in pesetas, how much more the Spanish purchaser had to pay.

12. An article is sold in England for £2 and in Spain for 288 pesetas. The English price is 20% higher than the Spanish price. Find the rate of exchange.

13. An Englishman, returning home from India, brought with him a carpet for which he had paid 5000 rupees. If the value of the rupee was 6p, find how much the carpet cost him.

14. When the exchange rate between England and Denmark was £1 = 14·40 kroner, find the value in Danish money of £59·75.

15. When £1 = 1·80 dollars, express, to the nearest penny, 475 dollars in English money.

CHAPTER 18

Foreign Exchange Calculations— the Chain Rule

The *Chain Rule* is an arrangement by means of which problems involving several proportion sums may be solved in one operation. It is a simple method of solving many exchange problems. The following are the steps:

1. State the question, putting ? in front of the missing term, *e.g.* Francs? = 1 kg.
2. Each line starts with the same denomination as the previous line ended.
3. Equivalents are placed opposite each other.
4. The last term is of the same denomination as the first, *i.e.* the missing term.
5. The missing term

$$= \frac{\text{Product of terms on complete side}}{\text{Product of terms on incomplete side}}.$$

Ex. 1   3 lb of cheese cost £2·25. Give an equivalent cost in francs per kilogramme, when 1 kg = 2·2 lb, and £1 = 10·80 francs.

How many francs? = 1 kg?
$$1 \text{ kg} = 2 \cdot 2 \text{ lb}$$
$$3 \text{ lb} = £2 \cdot 25$$
$$*100\text{p} = £1$$
$$£1 = 10 \cdot 80 \text{ francs}$$

$$\text{Equivalent price} = \frac{2 \cdot 2 \times 225 \times 10 \cdot 8}{3 \times 100}$$

$$= \textbf{17·82 francs per kg} \quad \textbf{Ans.}$$

* Any known equivalent may be inserted in order to get a required denomination. In this case, £1. 100p = £1.

Ex. 2   Given that 1 kg = 2·2 lb, and £1 = 11·00 francs, find the English price per lb equivalent to 27·50 francs per kg.

$$
\begin{array}{r}
\text{p?} \diagup \text{1 lb} \\
2·2 \text{ lb} \diagdown \text{1 kg} \\
\text{1 kg} \diagup 27·50 \text{ fr.} \\
11·00 \text{ fr.} \diagdown £1 \\
£1 \diagup 100\text{p}
\end{array}
$$

$$
\text{Equivalent price} = \frac{\overset{250}{\cancel{2750}}}{2·2 \times \cancel{11}}
$$

$$
= \textbf{£1·14} \qquad\qquad \textbf{Ans.}
$$

The chain rule may be used with advantage to solve any problem involving several proportionals.

Ex. 3   In two heats of a 100-metre race, A gives B 3 m start and beats him by 5 m. B gives C 2 m start and beats him by 3 m. C then runs against A, who gives him 8 m start. Which should win, and by how much?

$$
\begin{array}{r}
\text{C?} \diagup \text{A } 100 \\
\text{A } 100 \diagdown \text{B } 92 \\
\text{B } 100 \diagup \text{C } 95
\end{array}
$$

While A runs 100,

$$
\text{C runs } \frac{\overset{46}{\cancel{100}} \times 92 \times \overset{19}{95}}{\cancel{100} \times \cancel{100}} \text{ m}
\qquad
\begin{bmatrix} 460 \\ 414 \\ \overline{874} \end{bmatrix}
$$

$$
= 87·4 \text{ m.}
$$

C receives 8 m start, so A *wins* by 100 − 95·4 m

$$
= \textbf{4·6 m} \qquad\qquad \textbf{Ans.}
$$

Ex. 4   Give a multiplier for converting pence per lb to pesetas per kg, when £1 = 180 pesetas. (1 kg = 2·2 lb.)

(If we re-word this question to find an equivalent price to 1p for 1 lb, the equivalent will be our multiplier.)

$$\begin{array}{r} \text{pes?} \diagup \text{1 kg} \\ \text{1 kg} \diagdown \text{2·2 lb} \\ \text{1 lb} \diagup \text{1p} \\ \text{100p} \diagdown \text{£1} \\ \text{£1} \diagup \text{180 pes.} \end{array}$$

$$\text{Multiplier} = \frac{2\cdot2 \times 180}{100}$$

$$= \mathbf{3\cdot96} \qquad \textbf{Ans.}$$

Ex. 5   The distance between two Spanish towns is 104 km, and the fare is 110 pesetas. Give an equivalent fare from London to Brighton, which is 52 miles. Given that £1 = 180 pesetas. (8 km = 5 miles.)

$$\begin{array}{r} \text{£?} \diagup \text{52 miles} \\ \text{5 miles} \diagdown \text{8 km} \\ \text{104 km} \diagdown \text{110 pes.} \\ \text{180 pes.} \diagup \text{£1.} \end{array}$$

$$\text{Equivalent fare} = \pounds\frac{52 \times \overset{1}{\cancel{8}} \times \cancel{110}}{5 \times \underset{13}{\cancel{104}} \times \cancel{180}}$$

$$= \mathbf{49p} \text{ (to nearest p) Ans.}$$

Exercise 18

1. If sugar is 4·00 francs per kg in Paris, find, to the nearest penny, an equivalent price per lb in London, when £1 = 10·80 francs. (1 kg = 2·2 lb.)

2. The fare on a British railway is 10p a mile. Give an equivalent fare in francs for a journey of 25 km on a French railway, when £1 = 11·00 francs. (8 km = 5 miles.)

3. When £1 is worth 11·00 francs and 1 dollar is worth 6·00 francs, give the value of £1 in dollars and cents.

4. A liqueur is sold in Belgium at 236 francs a litre. Give an equivalent price in English money per pint, when the rate of exchange is 72 francs for £1. (1 l = 1·76 pt.)

5. An Englishman bought a watch in Belgium for 1500 francs, when the rate of exchange was 74 francs for £1. He paid import duty at 50% and VAT at 15% of the price paid *plus* import duty. What did the watch cost him in sterling?

6. If the value of 100 feet of brocade is £120, what is its value in pesetas per metre, to the nearest peseta? (£1 = 182 pesetas, 1 m = 39·37 in.)

7. In 2 heats of a 200-metre race, A gives B 7 m start and beats him by 8 m. B gives C 5 m start and loses by 10 m. C then runs against A who gives him 12 m. Which should win and by how much?

8. Find a multiplier, to 3 decimal places, for converting pence per foot to pesetas per metre, given that £1 = 181 pesetas, and 1 m = 39·37 in.

9. In Germany butter was sold at 11·00 marks per kilogramme. In England it was 98p a lb. If the rate of exchange was £1 = 4·60 marks, find how much cheaper it was in England. (1 kg = 2·2 lb.)

10. Give, in centimes per kg, an equivalent price to £96 per ton, given that £1 = 8·40 francs and 1 kg = 2·205 lb.

11. An Englishman bought 12 ha of land for 500 000 francs. Give the price of the land in £'s per acre, to the nearest £, when £1 = 11·00 francs. (1 hectare = 2·47 acres.)

12. When the Hong Kong dollar is worth 9p and the Australian dollar is 1·62 to the £ (sterling), find the value in Hong Kong dollars of the Australian dollar.

13. If $A.1 = 11·12 belga and 58 pesetas, find the value of 475 pesetas in francs and centimes. (1 belga = 5 francs.)

14. Material is sold in Paris at 10 francs a metre. Find the price per foot in dollars and cents to the nearest cent, when $A.1 = 5·48 francs. (1 m = 39·37 in.)

15. A plot of land is valued at $A.3200 an acre. Find its value in francs per square metre when $A.1 = 5·50 francs. (Take 1 m to be 39 in.)

16. When copper is quoted at $980 per ton in Australia, find, to the nearest dollar, an equivalent price per short ton in the U.S.A. when the rate of exchange is $U.S.1 = $A.0·8925.

CHAPTER 19

Rates and Taxes

1. Rates

Rates are the amounts charged by a Local Authority on the owners or occupiers of property in its area for the purpose of meeting the municipal expenditure.

Each house or other property in the area is assessed (*i.e.* its value is determined) at a certain amount, which is known as the rateable value. The total of these assessments is the rateable value of the area.

The rate is expressed as a certain amount in the £ of rateable value. In order to arrive at the rate, the estimated expenditure of the coming year is divided by the total rateable value

$$\text{Rate} = \frac{\text{Estimated expenditure}}{\text{Rateable value}}.$$

Rates are usually payable half-yearly or quarterly.

Ex. 1 The estimated expenditure of a certain borough is £1 740 000, and its rateable value is £3 000 000. Find the rate.

$$\text{Rate} = £\tfrac{174}{300} \text{ in the } £.$$
$$= \mathbf{58p} \qquad \textbf{Ans.}$$

Ex. 2 The rates are payable half-yearly in the above borough. What will be the half-yearly payment by the tenant of a house assessed at £215?

$$\text{Half-yearly payment} = \frac{58p \times 215}{2}$$
$$= 29p \times 215$$
$$= \textbf{£62·35} \qquad \textbf{Ans.}$$

Ex. 3 If it were decided to build a school in the above borough at a cost of £380 000, by how much would the rate be increased?

$$\text{Increase} = £\frac{380\ 000}{3\ 000\ 000}$$

$$= \textbf{13p} \qquad \textbf{Ans.}$$

$$\begin{bmatrix} 150)19(0\cdot127 \\ \underline{15} \\ 40 \\ \underline{30} \\ 100 \end{bmatrix}$$

Ex. 4 A tenant pays £95·76 when the rates are 63p in the £. What is the rateable value of his house?

$$\text{Rateable value} = £\frac{95\cdot76}{1} \times \frac{100}{63}$$

$$= £\frac{9576}{63}$$

$$= \textbf{£152} \qquad \textbf{Ans.}$$

2. Taxes

Taxes are the amounts collected by the Central Government to meet national expenditure.

Indirect Taxes are taxes paid at source and later passed on to the consumer, such as excise duties on wines, spirits, and beer, and customs duties on imported commodities. Value Added Tax (VAT) is another indirect tax and is levied on all goods and services. It is currently set at 15% (though some items are zero rated) and as far as the customer/consumer is concerned it may or may not be included in the stated price.

Direct Taxes consist mainly of licences, death duties, and income tax, and are paid directly to the government.

Income Tax, as its name implies, is a tax on a person's income. The greater the income, the greater the tax. Before the introduction of the 'pay as you earn' system of collection in April, 1944, it was the custom for the income-tax payer to pay the amount of his income tax directly to the Commissioners of Inland Revenue through the local Collector of Taxes.

With the introduction of P.A.Y.E. the onus of collection is placed upon the employer, who deducts the tax from the wages or salary of his employee.

Each employee should have a code number. This is arrived at by the tax office from the return made by the employee, showing details of the allowances claimed. The total of the allowances, less the last figure, will fix the code number. For total allowances of £2345 the code number will, therefore, be 234.

The code number of each employee is sent to the employer together with tax tables showing the amounts to be deducted from salary or wages.

The tax table shown is for 1981/82. It should be noted that not all the various types of allowances are shown.

Band of taxable income £	Rate of tax %	Tax on band £	Cumulative tax £
0–11250	30	3375	3 375
11251–13250	40	800	4 175
13251–16750	45	1575	5 750
16751–22250	50	2750	8 500
22251–27750	55	3025	11 525
over 27750	60		

Personal Allowances

	£
Married man	2145
Single person	1375
Wife's earned income relief	1375
Age relief—married couple	2895
Age relief—single or widowed person	1820
Dependent relative	100

The interest payable on a mortgage may also be deducted from gross income before calculating the tax due.

Ex. 5   How much tax will be paid by a single person whose gross income is £4300?

$$
\begin{array}{lr}
 & \pounds \\
\text{Income} & 4300 \\
\textit{Less} \text{ single person's allowance} & \underline{1375} \\
\text{Tax chargeable on} & 2925 \\
\end{array}
$$

at 30% = **£877·5   Ans.**

Ex. 6   What will be the monthly salary of a married man with an £8000 mortgage at 11% interest whose gross income is £6200?

$$
\begin{array}{lr}
 & \pounds \\
\text{Income (gross)} & 6200 \\
\textit{Less} \text{ mortgage interest £8000 at 11\%} & 880 \\
\textit{Less} \text{ married man's allowance} & \underline{2145} \\
\text{Tax chargeable on} & 3175 \\
\end{array}
$$

at 30%   £952·50

Net salary = £6200 − £952·50 = £5247·50

= **£437·30 per month**   Ans.

Ex. 7   An elderly couple receive a State pension of £2895, investment income of £500 and an annuity of £3000. How much tax will they pay per annum?

$$
\begin{array}{lr}
 & \pounds \\
\text{State pension} & 2895 \\
\text{Investment income} & 500 \\
\text{Annuity} & \underline{3000} \\
\text{Gross income} & 6395 \\
\textit{Less} \text{ aged couple relief} & \underline{2895} \\
\text{Tax chargeable on} & 3500 \\
\end{array}
$$

at 30% = **£1050**   Ans.

Note that the personal allowance is designed to cover almost exactly any State pension so that those living on that alone will not pay any tax.

Ex. 8   A married man with a gross income of £9500 has a £12 000 mortgage at 11¾% interest. He pays £1000 p.a. into a pension and supports his widowed mother. What is the amount of his monthly pay cheque?

	£
Gross salary	9500
Less mortgage interest £12 000 at $11\frac{3}{4}\%$	1410
Less married man's zllowance	2145
Less pension fund	1000
Less dependent relative allowance	100
Tax chargeable on	4845

at 30% = £1453·50

Net income = £9500 − £1453·50 = £8046·50

= **£670·54 per month** **Ans.**

Ex. 9 A man's income is £5000. His code number is 128. Calculate:

(*a*) the amount of tax payable

(*b*) the percentage of his income that is paid in tax.

	£
Income	5000
Less code allowance	1280
Tax chargeable on	3720

at 30% = **£1116** **Ans.** (*a*)

Percentage of income $= \dfrac{1116}{5000} \times 100$

= **22·32%** **Ans.** (*b*)

Ex. 10 A company director received a total of £29 000 in payments from his business last year. He pays £1200 p.a. into an annuity fund and has a £10 000 mortgage at 12%. How much tax will he pay?

	£
Gross income	29 000
Less annuity payments	1 200
Less mortgage £10 000 at 12%	1 200
Less married man's allowance	2 145
Tax chargeable on	£24 455

$$\text{Tax on } \pounds\text{0–11 250} \qquad \text{at } 30\% \qquad 3\ 375$$

Tax on £0–11 250	at 30%	3 375
£11 251–13 250	at 40%	800
£13 251–16 750	at 45%	1 575
£16 751–22 250	at 50%	2 750
£22 251–24 455	at 55%	1 212
∴ Total tax payable		**£9 712** Ans.

Ex. 11   In 1979 the VAT on furniture rose from 8% to 15%. How much more would be paid for a settee which before the change in the rate cost £135 including VAT?

First method: Increase in price $= \pounds\dfrac{135}{1} \times \dfrac{7}{108}$

$$= \pounds\mathbf{8 \cdot 75} \qquad\qquad \textbf{Ans.}$$

Second method: Price with increased VAT $= \pounds\dfrac{135 \times 115}{108}$

$$= \pounds 143 \cdot 75$$

$$\text{Increase} = \textbf{£8·75 Ans.}$$

Third method: Basic price $= \pounds\dfrac{135}{1} \times \dfrac{100}{108}$

$$= \pounds 125$$

Price with increased VAT $= \pounds\dfrac{125}{1} \times \dfrac{115}{100}$

$$= \pounds 143 \cdot 75$$

$$\therefore \text{Increase} = \textbf{£8·75} \qquad\qquad \textbf{Ans.}$$

Exercise 19

1. The rateable value of a town is £2 573 468. The estimated expenditure is £3 180 806. Find the rate to the nearest penny.

2. What will a tenant of a house in the above town pay in half-yearly rates if his house is assessed at £154?

3. It is decided to acquire a site and construct a community centre at a cost of £550 000. What will be the increase in the rate?

4. A man pays £80 per month rent. His rates are 94p in the £ payable on one-fifth of the rent. What is the total of his rent and rates per annum?

5. A man pays £480 a year rent and rates. If the rates are three-sevenths of the rent, how much does he pay monthly for rent?

6. If the rates in a borough are raised from 72p to 82p, how much more per annum will a man pay who is assessed on two-fifths of the value of his rent of £65 per month?

7. When the rate is £1·04, a tenant pays £102·96. What is the rateable value of his house?

8. In a certain borough, a rate of 88p brings in £2 323 332. Find the rateable value of the property in the borough.

The following questions all relate to the 1981/82 tax year.

How much income tax should be paid by persons having the following incomes:

9. Single person with an income of £3250?

10. Married man with an income of £3250?

11. A married couple when the man earns £3800 and the woman earns £2400?

12. A single woman earning £6500 per annum who supports her widowed father and who has a mortgage of £3000 at $11\frac{3}{4}\%$?

13. A labourer earns £85 per week and pays 5% into a retirement scheme. What is his tax code number assuming he is (*a*) single, (*b*) married?

14. A shipowner earned £20 500 last year. He is married and employs his wife as a secretary at a salary of £3000. His house is mortgaged for £16 000 at 12%. What is their joint net income? If his wife had decided not to work, by how much would this change their net income?

15. A single woman earns a weekly wage of £122. Her tax code is 195. How much will she receive in one year, net?

16. When the VAT rate went from 8% to 15% what

would be the new price of an item which previously cost £28·08?

17. What is the extra cost of an item previously costing £513 including VAT at 8% when the rate now changes to 15%?

CHAPTER 20

Averages and Mixtures

1. Averages

An *Average* or *Mean* is the quotient obtained by adding together a series of quantities and dividing the sum by the number of quantities.

Ex. 1 Find the average of 99, 107, 125, and 85.

$$\text{Average} = \frac{99 + 107 + 125 + 85}{4}$$

$$= \frac{416}{4} = \mathbf{104} \qquad \textbf{Ans.}$$

Ex. 2 The heights of 3 boys are 1·35 m, 1·42 m, and 1·52 m. Find the average height.

$$\text{Average height} = \frac{4·29}{3} \text{ m}$$

$$= \mathbf{1·43 \ m} \qquad \textbf{Ans.}$$

Ex. 3 Find the average age of 5 people whose ages are 24 years, 19 years, 35 years, 16 years, and 47 years.

$$\text{Average age} = \frac{141}{5}$$

$$= \mathbf{28·2 \ years} \qquad \textbf{Ans.}$$

Ex. 4 A man cycles 25 km at 15 km/h. Then he walks 5 km at 6 km/h. Find, his average speed for the whole journey.

$$25 \text{ km take } \tfrac{25}{15} \text{ h} = 1\tfrac{2}{3} \text{ h.}$$
$$5 \text{ ,, ,, } \tfrac{5}{6} \text{ ,, } = \tfrac{5}{6} \text{ ,,}$$
$$\overline{30 \text{ km take } \qquad\qquad 2\tfrac{1}{2} \text{ h.}}$$

$$\text{Speed} = \frac{30}{2\tfrac{1}{2}} \text{ km/h}$$
$$= \frac{60}{5} \text{ ,,}$$
$$= \textbf{12 km/h} \qquad\qquad \textbf{Ans.}$$

Ex. 5   A firm's sales from Jan. to May were £857, £942, £898, £935, £979. What must be the June sales in order to make the monthly average £940.

Total for 6 months must be £940 × 6 = £5640.

June sales must be £5640 − £4611

$$= \textbf{£1029} \qquad \textbf{Ans.}$$

$$\begin{bmatrix} 857 \\ 942 \\ 898 \\ 935 \\ 979 \\ \hline 4611 \end{bmatrix}$$

Ex. 6   The following are the weekly wages paid by a certain firm to junior employees.   25 receive £48 each, 50 receive £40 each, and 33 receive £36 each.   Find the average weekly wage per junior employee.

$$\begin{array}{lr} & £ \\ 25 \text{ men at } £48 = & 1200 \\ 50 \text{ men at } £40 = & 2000 \\ \underline{33 \text{ men at } £36 =} & \underline{1188} \\ \overline{108 \text{ men}} & 4388 \end{array}$$

$$\begin{bmatrix} 108)\overline{4388}(40\cdot629 \\ \quad\, \underline{432} \\ \qquad 680 \\ \qquad \underline{648} \\ \qquad\quad 320 \\ \qquad\quad \underline{216} \\ \qquad\qquad 104 \end{bmatrix}$$

$$\text{Average weekly wage} = £\frac{4388}{108}$$

$$= \textbf{£40·63} \qquad\qquad \textbf{Ans.}$$

2. Mixtures

Ex. 7   A wine merchant mixes 1 l of brandy, worth
£5·60, with 7 l of wine, worth £1·85 a litre, and calls the
mixture Branvin. What is a litre of Branvin worth?

$$
\begin{array}{ll}
1 \text{ l at } £5·60 = & 560\text{p} \\
\underline{7 \quad ,, \quad £1·85} = & \underline{1295\text{p}} \\
8 \text{ l} & 1855\text{p}
\end{array}
$$

1 l Branvin is worth $\frac{1855}{8}$p = **£2·32**   **Ans.**

Ex. 8   In what proportions should tea worth £2·66 a
kg be mixed with tea worth £2·94 a kg to produce a blend
worth £2·73 a kg?

Every kg of the first   is 7p below the average worth.
,,   ,,   second ,, 21p above   ,,   ,,
and 3 kg of the first   is 21p below the average worth.
,,   1 kg   ,,   second ,, 21p above   ,,   ,,

So a loss of 21p on the £2·66 tea is equated by a gain of
   21p on the £2·94 tea,
and the proportions are 3 kg at £2·66
                  1 kg at £2·94.

This may be stated as follows:

In pence

$$
\begin{array}{ccc}
266 & & 21 \\
& \searrow \quad \nearrow & \\
& 273 & \\
& \nearrow \quad \searrow & \\
294 & & 7
\end{array}
$$

Ratio = 3 : 1.
i.e. **3 kg at £2·66** $\Big\}$ **Ans.**
      1 kg at £2·94 $\Big\}$

Rules:
   1. Place the value of the ingredients in a column, with
      the average between them and to the right.
   2. Place the difference between the *first quantity* and
      the average opposite the second quantity.

3. Place the difference between the *second quantity* and the average opposite the first quantity.

4. This gives the ratio of the respective quantities.

Ex. 9   A man invested £1000 in stocks A and B. A paid a dividend of 6½%. B paid a dividend of 1½%. His average dividend on the two holdings was 5%. How much did he invest in each?

$$6\tfrac{1}{2} \qquad 3\tfrac{1}{2}$$

$$5$$

$$1\tfrac{1}{2} \qquad 1\tfrac{1}{2}$$

Ratio $= 3\tfrac{1}{2}:1\tfrac{1}{2} = 7:3$.

Investments $= \tfrac{7}{10}$ of £1000 $=$ **£700 in A** $\Big\}$ **Ans.**
                $\tfrac{3}{10}$,, $=$ **£300 in B**

Ex. 10   A man invested £1000 in two ventures. From *x* he gained 7%. From *y* he lost 3%. His average gain was 5%. How much did he invest in each?

$$+7 \qquad 8$$

$$5$$

$$-3 \qquad 2$$

Investments $= \tfrac{4}{5}$ of £1000 $=$ **£800 in** *x* $\Big\}$ **Ans.**
                $= \tfrac{1}{5}$,, $=$ **£200 in** *y*

Note: The difference between -3 and $+5 = 8$.

Exercise 20

1. Find the average of 87, 93, 45, and 55.

2. The heights of 4 men are 1·63 m, 1·75 m, 1·66 m, and 1·58 m. Find the average height.

3. A fruiterer sells 96 oranges at 8p each, 72 at 6p, and 120 at 5p each. Find, to the nearest new halfpenny, the average price for an orange.

4. The takings of a business for a year are £12 768. Counting 52 weeks to the year, 5½ days to the week, and allowing 6 days for bank holidays, what are the average daily takings?

5. Find the average age of 4 people, whose ages are 19 years, 43 years, 25 years, and 73 years.

6. A man cycles 36 km at 10 km/h. He then walks 7 km at 5 km/h. Find his average speed for the whole journey.

7. The following are the weekly wages paid by a firm: 18 men receive £100 each, 48 men receive £85 each, and 120 men receive £70 each. Find the average weekly wage per man.

8. A firm's sales for the first 5 months of the year were: £1278, £2196, £1856, £2347, £2317. What must be the amount of the June sales in order to make a monthly average of £2000?

9. A grocer mixes 21 kg of tea worth £2·52 a kg with 9 kg of tea worth £1·82 a kg. What will be the value of 1 kg of the blend?

10. A grocer mixes tea worth £2·59 a kg with tea worth £2·03 a kg in order to produce a blend worth £2·38 a kg. In what proportions were the teas mixed?

11. A grocer sells a quarter of his stock of tea at £2·45 a kg and the remainder at £2·31 a kg. What were the average price per kg?

12. 3 tonnes of coal are divided among 3 families consisting of 3, 4, and 5 persons respectively. Each person receives the same amount. How much did each family receive?

13. A man invested £600 in two ventures. From x he gained 7% and from y he gained 4%. His average gain was 5%. How much did he invest in each?

14. A man invested £800 in two ventures. From x he

gained 11% and from y he lost 5%. His average gain was 7%. How much did he invest in each?

15. A man's income for 12 months averaged £390 per month. In the first 3 months his average was £450 per month and in the next 4 months it was £300 per month. What was his average monthly income for the rest of the year?

CHAPTER 21

Commission—Brokerage—
Insurance—Bankruptcy

1. Commission

An *agent* is a person who acts on behalf of another person
called the *principal*.

When a person employs another person to buy or sell
property or goods of any kind, he usually pays the agent
by a percentage of the amount involved. This percentage
is called a *commission*.

Ex. 1   An auctioneer charges a commission of $7\frac{1}{2}\%$ on
goods sold. Jones sent a table to an auction and it was
sold for £76. How much would Jones receive?

$$\text{Amount} = £76 - 7\tfrac{1}{2}\% \text{ of } £76$$
$$= 76 - £5\cdot70$$
$$= £70\cdot30 \qquad \textbf{Ans.}$$

$$Or \qquad £\frac{76 \times 92\frac{1}{2}}{100}$$
$$= £70\cdot30 \qquad \textbf{Ans.}$$

Ex. 2   Smith received £$10\cdot17\frac{1}{2}$ for a gas fire which had
been sold at an auction sale. If the auctioneer charged a
commission of $7\frac{1}{2}$ per cent., for how much was the gas stove
sold?

$$\text{Selling price} = £\frac{10\cdot17\frac{1}{2}}{1} \times \frac{100}{92\frac{1}{2}}$$
$$= £\frac{2035}{185}$$
$$= £11 \qquad \textbf{Ans.}$$

Ex. 3   A traveller receives a salary of £1500 and a commission of 10% on all orders over the first £1000. If his sales last year were £29 500, what would be the amount of his income?

$$\text{Income} = 10\% \text{ of } £28\,500 + £1500$$
$$= £2850 + £1500$$
$$= \textbf{£4350} \qquad \textbf{Ans.}$$

Ex. 4   An estate agent charges for the sale of property, 5% on the first £1000 and $2\frac{1}{2}\%$ on the remainder. If the vendor (seller) received £25 325, how much did the purchaser pay for a house?

On 1st £1000 vendor receives £950.

$$£25\,325 - £950 = £24\,375.$$

Vendor receives £97·50 on each £100.

$$\text{Purchaser paid } £\frac{24\,375}{1} \times \frac{100}{97\cdot50} + £300$$
$$= £25\,000 + £1000$$
$$= \textbf{£26\,000} \qquad \textbf{Ans.}$$

2. Brokerage

Brokerage is a commission charged by a broker for the purchase or sale of stocks and shares. It is reckoned as a percentage on the sum, called the *consideration*, for which they are bought or sold.

Nominal value is the face (or named) value.

Par is the term used to denote a price which coincides with the nominal value.

Premium denotes a higher price than nominal value.

Discount denotes a lower price than nominal value.

Ex. 5   A man bought £500 of stock quoted at 96, brokerage $1\frac{1}{4}\%$. What was his outlay?

$$\text{Consideration} = £\frac{500 \times 96}{100} = £480$$

$$\text{Brokerage} = £\frac{450 \times 1\frac{1}{4}}{100} = \quad 6$$

$$\text{Outlay} = \qquad\qquad \underline{\textbf{£486}} \qquad \textbf{Ans.}$$

Ex. 6   A man sold 800 £1 shares, quoted at 76p. If the brokerage was $1\frac{1}{4}$ per cent., how much did he receive?

$$\text{Consideration} = 76\text{p} \times 800 = £608$$

$$\text{Commission} = £\frac{608 \times 1\frac{1}{4}}{100}$$

$$= £\frac{608}{80} \qquad = \qquad 7\cdot60$$

$$\text{Amount received} = \qquad \underline{\textbf{£600}\cdot\textbf{40}} \quad \textbf{Ans.}$$

3. Insurance

The *Policy* is the contract between the person or firm taking out the insurance and the insurance company.

The *Premium* is the weekly, monthly, quarterly, half-yearly, or yearly payment.

Ex. 7   A man insures the contents of his house at a premium of 25p per £100. How much will he pay each half-year if he values the contents at £6000?

$$\text{Amount} = \tfrac{1}{2} \text{ of } \tfrac{1}{4}\% \text{ of } £6000$$
$$= £\frac{6000}{800}$$
$$= \textbf{£7}\cdot\textbf{50} \qquad\qquad \textbf{Ans.}$$

Ex. 8   A boat is valued at £45 000 and is insured for three-fifths of its value at a premium of 3%. What will be the amount of the premium?

$$\text{Premium} = £\frac{45\,000}{1} \times \tfrac{3}{5} \times \tfrac{3}{100}$$
$$= \textbf{£810} \qquad\qquad \textbf{Ans.}$$

4. Bankruptcy

Assets are the things a person possesses.

Liabilities are the amounts a person owes.

A person is said to be *insolvent* and may be declared a *bankrupt* when his liabilities are greater than his assets.

The amount which a bankrupt is able to pay his creditors is called a *dividend*.

The dividend is usually expressed as a certain sum in the £.

$$\text{The dividend} = \frac{\text{Assets}}{\text{Liabilities}}.$$

Ex. 9   A bankrupt's assets realise £6750 and his liabilities are £7500. Find: (*a*) the amount of the dividend, and (*b*) the sum a creditor for £400 will receive.

$$\text{Dividend} = £\tfrac{6750}{7500}$$
$$= \textbf{90p in the £} \qquad \textbf{Ans.} \ (a)$$
$$\text{Sum} = 90\text{p} \times 400$$
$$= \textbf{£360} \qquad \textbf{Ans.} \ (b)$$

Ex. 10   A bankrupt's assets realise £1875, and he pays 75p in the £. What were his liabilities?

$$\text{Liabilities} = £\tfrac{1875}{1} \times \tfrac{100}{75}$$
$$= \textbf{£2500} \qquad \textbf{Ans.}$$

Exercise 21

1. A traveller receives $7\frac{1}{2}\%$ commission on all sales above the first £500. If his sales amount to £15 700, what is his commission?

2. An auctioneer charges a commission of $7\frac{1}{2}\%$ on goods sold. How much should be received for a table which he sold for £84?

3. Jones received £24·97½ for a divan which had been sold by auction. The auctioneer had deducted $7\frac{1}{2}\%$ commission. For how much was the divan sold?

4. A traveller is paid £25 a week and a yearly commission of 10% on sales in excess of a monthly average of £500. His sales last year amounted to £65 300. What was his total income?

5. An author is offered an immediate payment of £360 for a book, or a royalty of 3p on each copy sold. How many books must be sold in order to equate these offers?

6. A salesman is offered a wage of £80 a week, or a commission of 8% on all goods sold. What must be his sales for a year (52 weeks) in order to equate these offers?

7. A traveller receives a salary of £1500 a year and a commission of 8% on all sales over the first £5000. His total salary and commission last year was £5348. What were his sales?

8. An estate agent sold a house for £32 500. He charges a commission of 5% on the first £1000 and 2% on the remainder. How much should the owner receive?

9. A man bought £480 of stock quoted at $57\frac{1}{4}$. Brokerage was $1\frac{1}{4}$ per cent. How much did the stock cost him?

10. A man invested £500 in stock quoted at 80. Brokerage was $1\frac{1}{4}$ per cent. How much stock did he receive?

11. A man bought 180 £1 shares when their market price was 64p. Brokerage was $1\frac{1}{4}$%. What was his outlay?

12. £750 of stock was sold at $43\frac{1}{2}$. Brokerage was $1\frac{1}{4}$ per cent. How much was received for the stock?

13. £450 was received for stock which had been sold for 56. Brokerage was $1\frac{1}{4}$ per cent. What was the nominal value of the stock?

14. A man insures his life for £1000, and pays a half-yearly premium of £24·45. What percentage is the annual premium of the sum insured?

15. What will be the amount of the premium for insuring a vessel and cargo valued at £576 750 at $3\frac{3}{4}$%?

16. A man insures the contents of his house, valued at £5600, at a premium of 25p per £100. What premium will he pay?

17. A yacht worth £65 000 is insured for four-fifths of its value at a premium of 1¾%. What will be the amount of the premium?

18. A bankrupt's assets are £1545 and his liabilities are £3600. What dividend can he pay?

19. A creditor for £450 received a dividend of £348·75 from a bankrupt's estate. How much was this in the £?

20. A bankrupt's net assets were £4450, and he can pay a dividend of 37p in the £. What were his liabilities?

CHAPTER 22

Bills of Exchange—Discounting

A *Bill of Exchange* is defined as 'an unconditional order in writing, addressed by one person to another, signed by the person giving it, requiring the person to whom it is addressed to pay on demand, or at a fixed or determinable future time, a sum certain in money to or to the order of a specified person or to bearer'.

Expressed briefly, it is an order to a person to pay a specified sum of money on a given date.

A *Cheque* is a bill of exchange drawn on a banker and payable on demand.

The following is an example of a bill of exchange:

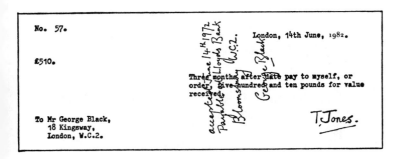

After a bill has been accepted the drawer may:

(1) Keep it until due, and then present it for payment.
(2) Endorse it and use it in payment of a debt, *i.e.* transfer the bill by endorsement.
(3) Sell it to a bank. This is usually referred to as 'discounting' the bill.

The due date of a bill is the number of days or months for which it is drawn *plus* 3 days of grace. The due date of the above bill would be 17th September, 1982.

Banker's discount is the Simple Interest at the agreed rate for the number of days the bill has to run at the date it is discounted.

Discountable value is the face value of the bill *minus* the discount.

In counting days, omit the first and count the last. If, for example, a bill was discounted on the 10th July, subtract 10 from 31, which gives 21 days for July and then add to this all the remaining days, including the day it is due.

If the above bill was discounted on the 20th June, the interest days would be:

> 10 in June
> 31 ,, July
> 31 ,, August
> 17 ,, September

Total = 89 days.

The due date of a 60-day bill drawn on the 5th July would be $(26 + 31 + 6 = 63)$ 6th September.

Ex. 1   Find the discountable value of a bill for £350, drawn on the 3rd May at 3 months and discounted on the 25th May at 9%.

Due date   = 6th August.

Days to run $= 6 + 30 + 31 + 6 = 73$.

Discount $= £\dfrac{350 \times 9 \times \overset{73}{\cancel{73}}}{100 \times \underset{5}{\cancel{365}}}$

$= £6 \cdot 30$

Discountable value $= £350 - £6 \cdot 30$

$= £343 \cdot 70$ **Ans.**

Ex. 2   Find the banker's discount on a 60-day bill for £450 drawn on the 10th May and discounted on the 4th June at 5%.

Days run   = 25.

Days to run = 63 − 25 = 38.

Discount   = £$\dfrac{450 \times 5 \times 38}{100 \times 365}$

            = **£2·34**   **Ans.**

$$\begin{array}{r} 2\text{·}342 \\ 73)171(\\ 146 \\ \hline 250 \\ 219 \\ \hline 310 \\ 292 \\ \hline 180 \end{array}$$

The *Present Worth*, or present value, is a sum of money which, invested at a given rate, will produce the given amount in the given time. Thus, £500 at 6% per annum will amount to £507·50 in 3 months, and we may say that the present worth of £507·50 due 3 months hence at 6% per annum is £500.

True Discount is the difference between the present worth and the amount. It is the simple interest on the sum received. Banker's Discount is simple interest on the amount while True Discount is simple interest on the principal.

Ex. 3   Find the present value of £1000 due 5 months hence at 6% per annum.

Interest on £1 = £$\dfrac{1 \times 5 \times 6}{100 \times 12}$ = £$\dfrac{1}{40}$.

Amount of £1 = £$1\frac{1}{40}$ = £$\frac{41}{40}$.

(For each £$\frac{41}{40}$ in £1000 the present value is £1, and by dividing £1000 by $\frac{41}{40}$ we get the present value of £1000.)

∴ Present value of £1000

  = £$\frac{1000}{1}$ ÷ $\frac{41}{40}$

  = £$\frac{1000}{1}$ × $\frac{41}{40}$

  = **£975·61**   **Ans.**

$$\begin{array}{r} 975\text{·}609 \\ 41)40000(\\ 369 \\ \hline 310 \\ 287 \\ \hline 230 \\ 205 \\ \hline 250 \\ 246 \\ \hline 40 \end{array}$$

Ex. 4 Find the difference between the true discount and the banker's discount at 5% on a bill of exchange for £500 which has 73 days to run.

Banker's discount $= £\dfrac{500 \times 73 \times 5}{365 \times 100}$

$= £5.$

Present worth $= £\dfrac{500}{1} \times \dfrac{500}{505}$

$= £495 \cdot 05.$

True discount $= £500 - £495 \cdot 05$

$= £4 \cdot 95.$

Difference $= £5 - £4 \cdot 95$

$= \textbf{5p}$ **Ans.**

$$\left[\begin{array}{r} 495 \cdot 0495 \\ 101)50000(\\ \underline{404} \\ 960 \\ \underline{909} \\ 510 \\ \underline{505} \\ 50 \end{array}\right]$$

Exercise 22

1. Find the discountable value of a bill for £292, drawn on the 5th March at 3 months and discounted on the 7th April at 6¾%.

2. Find the discountable value of a 60-day bill for £360, drawn on the 11th June and discounted on the 14th June at 9%.

3. Find the banker's discount on a 90-day bill of £450, drawn on the 5th November and discounted on the 8th December at 12½%.

4. Find the banker's discount on a bill of £438, drawn on the 10th July at 6 months and discounted on the 5th September at 11¼%.

5. How much will a man receive for a bill of £480, drawn on the 3rd May at 3 months and discounted on the 25th May at 9¾%?

6. Find the present worth of £650 due in 219 days, at 7½%.

7. Find the present worth of £350 due in 5 months, at 4%.

8. What sum will amount to £120 in 4 months at 3%?

9. What is the difference between the true and banker's discount on a 6 months' bill of £508·50, drawn on the 15th March and discounted on the 25th April at 4½%?

10. What is the difference between the true and banker's discount on a bill of £876 for 6 months, drawn on the 4th March and discounted on the 10th April at 5½%?

CHAPTER 23

Problems in Speed and Distance

A *speed* is the distance covered in a certain unit of time. It may be expressed in:

      1. Kilometres per hour (km/h).
      2. Metres per second (m/s).

$$1.\ \text{Speed} = \frac{\text{Distance in kilometres}}{\text{Time in hours}}\ \text{km/h.}$$

$$2.\ \text{Speed} = \frac{\text{Distance in metres}}{\text{Time in seconds}}\ \text{m/s.}$$

Ex. 1 A car travels 100 km in 1 h 40 min. Find its speed in kilometres per hour.

$$\text{Speed} = \frac{100}{1\frac{2}{3}}\ \text{km/h}$$

$$= \frac{300}{5}\quad\text{,,}$$

$$= \textbf{60 km/h}\qquad\qquad\textbf{Ans.}$$

Ex. 2 An aeroplane travels 480 km in 1 h 12 min. Find its speed in metres per second.

$$\text{Speed} = \frac{480 \times 1000}{1\frac{1}{5} \times 3600}\ \text{m/s}$$

$$= \frac{\overset{20}{\cancel{480}} \times 5 \times 10}{\underset{3}{\cancel{6}} \times \underset{3}{\cancel{36}}}$$

$$= \textbf{111 m/s}\qquad\qquad\textbf{Ans.}$$

Ex. 3 An aeroplane travels 650 km in 2 h 10 min. Give its speed: (*a*) in kilometres per hour, and (*b*) in metres per second.

$$\text{Speed} = \frac{650}{2\frac{1}{6}} \text{ km/h}$$

$$= \frac{650 \times 6}{13} \text{ km/h}$$

$$= \textbf{300 km/h} \qquad \text{Ans. } (a)$$

$$= \frac{\cancel{300} \times 1000}{\underset{12}{\cancel{3600}}} \text{ m/s}$$

$$= \textbf{83 m/s} \qquad \text{Ans. } (b)$$

Note: 1 km = 1000 m. 1 h = 3600 s.

Ratio = 1000 : 3600 = 5 : 18.

To convert kilometres per hour to metres per second:
Multiply by $\frac{5}{18}$.

To convert metres per second to kilometres per hour:
Multiply by $\frac{18}{5}$.

Ex. 4 Convert a speed of 36 km/h to metres per second.

$$\text{Speed} = \frac{36}{1} \times \frac{5}{18} \text{ m/s}$$

$$= \textbf{10 m/s} \qquad \text{Ans.}$$

Ex. 5 Sound travels at a speed of 335 m/s. Convert this to kilometres per hour.

$$\text{Speed} = \frac{335 \times 18}{5} \text{ km/h}$$

$$= \textbf{1206 km/h} \qquad \text{Ans.}$$

To find time taken to travel 1 km:

$$\text{Time} = \frac{\text{Time in hours}}{\text{Distance in kilometres}} \text{ hours.}$$

Or
$$\qquad \text{,,} \quad = \frac{\text{Time in minutes}}{\text{Distance in kilometres}} \text{ minutes.}$$

Ex. 6   A cyclist travels 45 km in 2 h 15 min. How long does he take to travel 1 km?

$$\text{Time} = \frac{2\frac{1}{4}}{45}\text{ h}$$

$$= \frac{9 \times 60}{4 \times 45}\text{ min}$$

$$= \textbf{3 min} \qquad\qquad \textbf{Ans.}$$

When two bodies are approaching each other:

$$\text{Time taken to meet} = \frac{\text{Distance apart in kilometres}}{\text{Sum of speeds in km/h}}\text{ hours.}$$

$$Or \quad\text{,,}\qquad\text{,,}\quad = \frac{\text{Distance apart in metres}}{\text{Sum of speeds in m/s}}\text{ seconds.}$$

Ex. 7   A and B are two towns 21 km apart. At 10 a.m. a cyclist starts from A to cycle to B at 10 km/h and a pedestrian starts from B to walk to A at 4 km/h. When and where will they meet?

$$\text{Distance apart} = 21\text{ km.}$$

$$\text{Time to meet} = \frac{21}{10 + 4}\text{h}$$

$$= 1\tfrac{1}{2}\text{ h.}$$

$$\therefore \text{ Time they meet} = \textbf{11.30 a.m.}\quad \textbf{Ans. (i)}$$

In $1\frac{1}{2}$ h cyclist has travelled:

$$1\tfrac{1}{2} \times 10\text{ km} = \textbf{15 km from A} \qquad \textbf{Ans. (ii)}$$

Note: We can check this by finding the pedestrian's distance from B. $1\frac{1}{2} \times 4 = 6$ km from B.

When two bodies are moving in the same direction:

$$\text{Time for faster to overtake slower} = \frac{\text{Distance apart}}{\text{Difference of speeds}}.$$

Ex. 8   A man starts from home to walk to the station 3 km away, at 8.30 a.m. He walks at 4 km/h. 10 min later, his wife finds he has forgotten his season ticket and immediately sets out on her bicycle, at 12 km/h. When and where will she overtake him?

> At 8.40, the man has walked $\frac{2}{3}$ km.
>
> Time taken to overtake him $= \dfrac{\frac{2}{3}}{12 - 4}$ h
>
> $\qquad\qquad\qquad\qquad = \dfrac{2 \times 60}{3 \times 8}$ min
>
> $\qquad\qquad\qquad\qquad = 5$ min.

Time $\quad = $ **8.45 a.m.**   **Ans.** (i)

$\therefore$ Place $= $ **1 km from home**   **Ans.** (ii)

Ex. 9   A and B are two towns 20 km apart. At 9.0 a.m. a motorist starts out from A for B at 30 km/h. He stays at B 20 min and then sets out to return to A. At 9.30 a.m. a pedestrian starts from A to walk to B at 5 km/h. When and where will they meet?

> Motorist arrives at B in 40 min.
>   ,,   stays 20 min.
> He starts back at 10 a.m.
> At 10 a.m. pedestrian has walked $2\frac{1}{2}$ km.
>
> Distance apart $\qquad = 17\frac{1}{2}$ km.
>
> Time taken to meet $\quad = \dfrac{17\frac{1}{2}}{30 + 5}$ h
>
> $\qquad\qquad\qquad\qquad = \dfrac{17\frac{1}{2}}{35}$ h
>
> $\qquad\qquad\qquad\qquad = \frac{1}{2}$ h.

$\therefore$ They meet at **10.30 a.m.**   **Ans.** (i)

Place $= $ **5 km from A**   **Ans.** (ii)

Check:
Motorist does 15 km in $\frac{1}{2}$ h. $\therefore$ He is 15 km from B.

Ex. 10   An aeroplane completes a course of 1000 km in 1 h $52\frac{1}{2}$ min. Find its speed in miles per hour. (8 km = 5 miles.)

$$\text{Speed} = \frac{1000}{1\frac{7}{8}} \times \frac{5}{8}\text{miles/h}$$

$$= \frac{1000 \times 8 \times 5}{15 \times 8} \text{ miles/h}$$

$$= \textbf{333}\tfrac{1}{3} \textbf{ miles/h} \qquad \textbf{Ans.}$$

Ex. 11   A man travels 60 km by train at 40 km/h, 15 km by cycle at 10 km/h, and he walks the last 9 km at 4 km/h. Find his average speed for the whole journey.

$$
\begin{array}{llll}
60 \text{ km at } 40 \text{ km/h} & = & 1\tfrac{1}{2} \text{ h} \\
15 \quad,, \quad 10 \quad ,, & = & 1\tfrac{1}{2} \text{ ,,} \\
\underline{9} \quad,, \quad 4 \quad ,, & = & \underline{2\tfrac{1}{4}} \text{ ,,} \\
84 \text{ km takes} & & 5\tfrac{1}{4} \text{ h.}
\end{array}
$$

$$\text{Average speed} \quad = \frac{84}{5\frac{1}{4}} \text{ km/h}$$

$$= \textbf{16 km/h} \qquad \textbf{Ans.}$$

Ex. 12   A cyclist rides from London to Crawley (52 km). The first half of the distance he does at 30 km/h. The second half at 20 km/h. Find his average speed for the whole journey.

$$26 \text{ km at } 30 \text{ km/h} = \frac{26}{30} \text{ h}$$

$$26 \quad ,, \quad 20 \quad ,, \quad = \frac{26}{20} \text{ ,,}$$

$$\text{Time for 52 km} \quad = \frac{26}{30} + \frac{26}{20} \text{ h}$$

$$= \frac{52 + 78}{60} \text{ ,,}$$

$$= \frac{13}{6} \text{ h.}$$

$$\therefore \text{ Speed} \quad = \frac{52}{1} \div \frac{13}{6} \text{ km/h}$$

$$= \frac{52}{1} \times \frac{6}{13} \quad ,,$$

$$= \textbf{24 km/h} \qquad \textbf{Ans.}$$

Exercise 23

1. A car travels 189 km in 2¼ h. Find its speed.

2. An aeroplane travels 1050 km in 3¾ h. Find its speed in miles per hour. (8 km = 5 miles.)

3. A man walks 108 km in 11 h 15 min. Find his speed in miles per hour.

4. An aeroplane travels 500 miles in 1 h 20 min. Find its speed in kilometres per hour.

5. An aeroplane travels 450 km in 1 h 52½ min. Give its speed in metres per second.

6. A car travels 250 km in 2 h 5 min. Find its speed: (*a*) in kilometres per hour, (*b*) in metres per second.

7. A motorist travelled 270 km in 3 h 45 min. How long did he take to travel 1 km.

8. A passenger in a train notices that it takes 1¼ s from one telegraph pole to the next. If the distance between telegraph poles is 44 m, what is the speed of the train?

9. A train left station A at 7.17, and arrived at station B, 31½ km distant, at 7.49. Find its speed to 0·01 km/h.

10. At 9.18 p.m. a burglar leaves a house, walking at 5 km/h. At 9.30 p.m. a policeman sets out after him on a cycle, riding at 11 km/h. Where and when will the policeman catch the burglar?

11. A cyclist travels from London to Crawley (52 km). He covers half the distance at 25 km/h and half at 20 km/h. Find his average speed for the whole journey.

12. A and B are two towns 30 km apart. At 10 a.m. a cyclist starts from A to cycle to B at 10 km/h. At 11.30 a.m. a motorist starts from B to motor to A at 30 km/h. When and where will they meet?

13. A cyclist makes a journey of 52 km. The first 20

km he travels at 10 km/h. He then rests for 30 min, and completes the rest of the journey at 8 km/h. What is his average speed for the whole journey?

14. A man does a journey of 75 km in a travelling time of 1 h. He travels 60 km by train and 15 km by car. The speed of the car is $\frac{3}{4}$ the speed of the train. What is the speed of the train?

15. A man makes a journey of 65 km. He travels 45 km by train at 30 km/h, 16 km by car at 32 km/h, and walks the last 4 km at 3 km/h. Find his average speed for the whole journey.

16. A boy runs 200 m in 36 s. Express this speed in kilometres per hour.

CHAPTER 24

Graphs

When two quantities are related in any way, the relationship is often brought out more clearly by means of a graph. Graphical representation of statistical data is of great importance in commerce and industry. The object of a graph is to convey information rapidly. The essentials of a good graph are *Neatness* and *Accuracy*.

When constructing a graph, the following points should be borne in mind:

1. The *scale* will be determined by the highest and lowest magnitudes. It should be as large as possible.
2. The *points* representing the simultaneous values should be placed very carefully.
3. The *plotting* or graphing (*i.e.* the joining of the points) should be done with very great care and neatness.
4. The scales and identity of the quantities should be indicated distinctly.
5. It is usual to represent the fixed quantity along the horizontal line (*OX*) and the variable quantity along the vertical line (*OY*). (See Fig. 1.)

Ex. 1   The following were a tradesman's sales during the first 6 months of the year. Construct a graph from the figures given.

January	£437.
February	£419.
March	£470.
April	£448.
May	£442.
June	£475.

The graph contains the same information as the column of figures, but in a much clearer form. (*See* Fig. 1.)

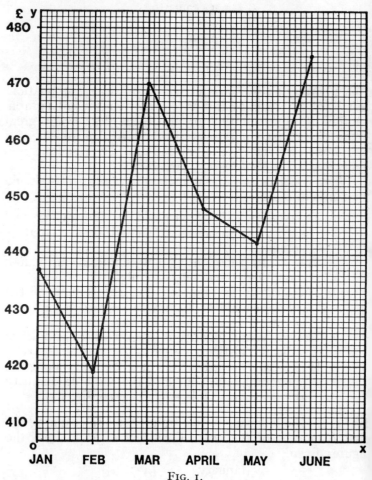

FIG. 1.

1. The salient characteristics are brought out.
2. The rate of increase or decrease is more apparent from the steep or gentle slope of the lines connecting the points.

The graph could be carried on from year to year on the same scale and axis by the use of different-coloured inks, and would form a quick method of comparing the sales of any particular month with those of the same months of previous years.

Interpolation is the process of finding intermediate values between the simultaneous values given. *Extrapolation* is the process of finding values *outside* those given and for this the graph must be continued beyond the plotted points. This may be done to *either* end of the graph.

Ex. 2   A boy earns £32 for a 40-hour week. Plot a graph to show the relationship between hours worked, up to 56, and money earned. From your graph find his earnings for a week in which he works: (*a*) 54 hours, (*b*) 36 hours, and (*c*) the number of hours he works in a week in which his wages are £20.

(Fig. 2) By interpolation we get points:

(*a*)	**£43·20**	**Ans.**	(*a*)
(*b*)	**£28·80**	**Ans.**	(*b*)
(*c*)	**25 hrs**	**Ans.**	(*c*)

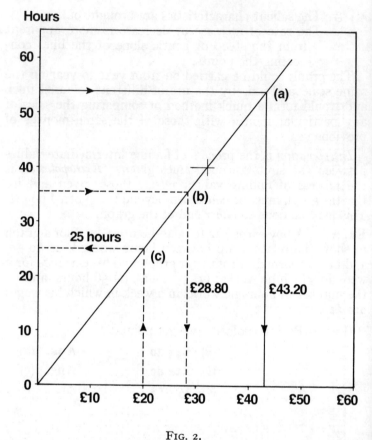

FIG. 2.

Ex. 3 Given that 1 kg = 2·2 lb, plot a graph for converting kg to lb up to 50 lb, and vice versa. From your graph convert: (*a*) 25 lb to kg, (*b*) 18·5 kg to lb, and (*c*) 15 kg to lb.

Note: It is a help in plotting a graph to obtain integral equivalents where possible. In this case we take 5 kg = 11 lb, or 10 kg = 22 lb.

(Fig. 3) By interpolation we get points:

$\qquad$ (a) **11·4 kg** $\qquad$ **Ans.** (a)

$\qquad$ (b) **40·7 lb** $\qquad$ **Ans.** (b)

$\qquad$ (c) **33 lb** $\qquad$ **Ans.** (c)

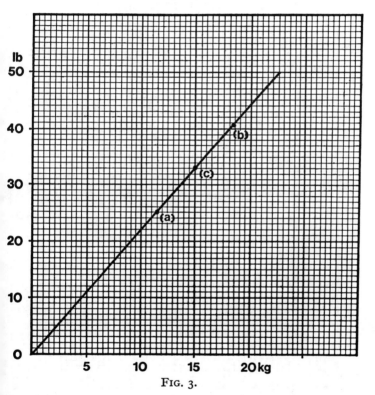

FIG. 3.

x. 4 A man made a journey of 35 km by train at
:o km/h, 13 km by car at 26 km/h, and he walked the last
▪ km at 3 km/h. He takes 15 min in the change from train
▪o car and 30 min before starting to walk. Construct a
▪ravel graph of the journey and find his average speed for
▪he whole journey.

(Fig. 4) A straight line (the broken line) from the start

to the end of the journey gives the average speed. After
1 h (point (*a*)) this line indicates 12 km.

∴ Average speed = **12 km/h**   **Ans.**

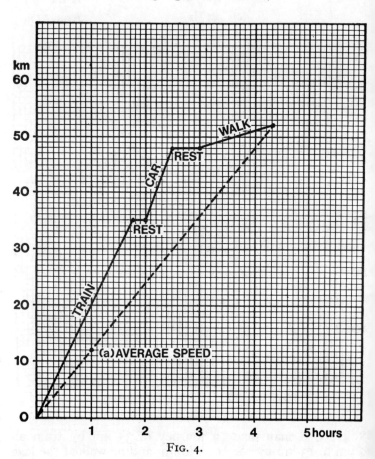

FIG. 4.

Ex. 5   Check, by means of a graph, the answer to Ex. 8,
Chapter 23. (*See* Fig. 5.)

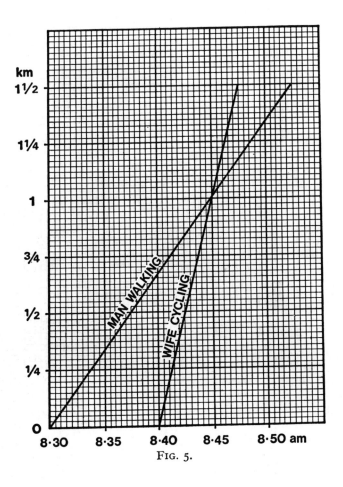

FIG. 5.

Ex. 6   Check, by means of a graph, the answer to Ex. 7,
Chapter 23.   (*See* Fig. 6.)

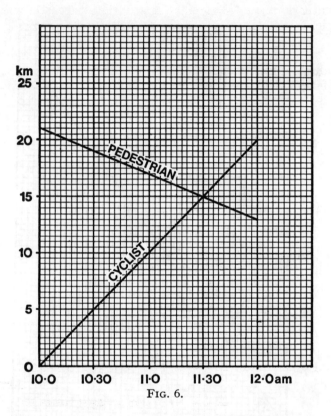

FIG. 6.

There are many forms of pictorial representation of data to be found in newspapers, magazines, on posters, and in Government publications. One of the methods of representing statistical data is the sector graph, in which a quantity is represented as the appropriate part (sector) of a circle.

Ex. 7 The following sector graph represents the units of production of four factories, A, B, C, and D.

If the total production of the four factories was 19 080 units, find the units of production of each factory.

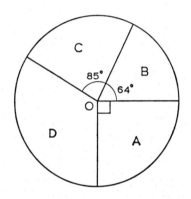

$$A \text{ (rt. angle)} = \frac{90}{360} \times 19\ 080 = \textbf{4770 units}$$

$$B = \frac{64}{360} \times 19\ 080 = \textbf{3392 units}$$

$$C = \frac{85}{360} \times 19\ 080 = \textbf{4505 units}$$

$$D\ (360°-239°) = \frac{121}{360} \times 19\ 080 = \textbf{6413 units}$$

Ans.

Note: The angle subtended at the centre of a circle is 360 degrees.

A sector of a circle is bounded by two radii and the arc which is a portion of the circumference.

The ratio of the area of a sector to the area of the circle, therefore, is:

$$\frac{\text{Angle of sector}}{360°} \times \text{Area of circle}$$

Another form of displaying statistical data is the bar chart, an example of which is shown below.

A merchant noted that sales of a particular item were as follows:

January	3	July	18
February	5	August	17
March	15	September	9
April	10	October	12
May	15	November	6
June	21	December	0

Draw a statistical chart to display this data.

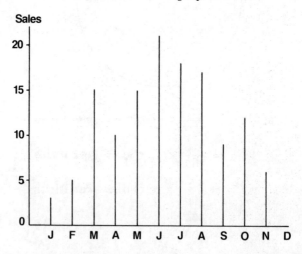

By observing the data displayed in this way it is possible to come to some conclusions regarding it. In the example shown, the merchant might not consider it worthwhile stocking this item during the winter months, for example.

Exercise 24

1. The following were a draper's sales during the first 6 months of the year. Plot a graph from the figures given: January, £588; February, £432; March, £500; April, £612; May, £540; June, £596.

Now make a bar chart to indicate the same information. Choose your scale carefully to emphasise the variation in sales. (Hint: start your vertical scale at £400.)

2. Given that 1 l = 1·76 pt, plot a graph for converting pints to litres and vice versa. From your graph convert (*a*) 22 pt to litres, and (*b*) 15 l to pints.

3. A workman earns £90 for a 40-hour week. Plot a graph showing the relationship between hours worked and money earned. From your graph find: (*a*) his wages for a 52-hour week, and (*b*) the hours worked when his wages were £99.

4. Plot a graph for converting miles to kilometres and vice versa, given that 8 km = 5 miles. From your graph convert: (*a*) 24 miles to kilometres, and (*b*) 36 kilometres to miles.

5. A traveller makes a journey of 30 km by train at 24 km/h, 18 km by bus at 12 km/h, and walks the remaining 7 km at 4 km/h. At each change he rests for 15 min. Find, by means of a travel graph, his average speed for the whole journey.

6. Check, by means of graphs, the answers to each of the following:

    (*a*) Ex. 9, Chapter 23.
    (*b*) Ex. 11, Chapter 23.
    (*c*) No. 11, Exercise 23.
    (*d*) ,, 12,   ,,
    (*e*) ,, 13,   ,,
    (*f*) ,, 15,   ,,

7. The following diagram shows the sales by a publisher of three books, A, B, and C. The total sales of these three

books was 38 160. How many copies of each book were sold?

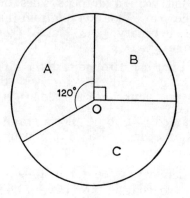

Answers

Exercise 1

1. A.   £

      2 034·18
      1 486·23
      1 545·73
      3 124·77
      2 223·20
      2 037·93
      3 704·00
        311·32
      ————
   £16 467·36

   £4 981·34
   £4 952·52
   £6 533·50

B.   £

      2 549·98
      2 665·61
      2 001·10
      3 245·85
      2 568·46
      5 227·40
      1 854·13
      1 087·08
      1 477·30
      2 412·56
      1 581·01
      1 595·35
      ————
   £28 265·83

   £9 272·95
   £7 648·35
   £11 344·53

2. A. Total £8027·56. Balance £310·31.
   B. ,,   £6470·91.   ,,   £148·13.
3. A. Total £171·44. Balance £9·55.
   B. ,,   £180·97.   ,,   £7·52.
4. Total £173·85. Cr. Balance £87·60.
5.   ,,   £95·97. Dr. ,,   £28·75.
6. A. 28·386.   B. 69 521.   C. 4 101 970.
7. A. 496.   B. 599.   C. 99.
8. 3·6 t.   9. 67.   10. £98·40.
11. £58 802·50.   12. 73.   13. 4854·8 fr.
14. £562 120.   15. 766 467 rupees.   16. $11 464 285.

Exercise 2

1. $\frac{9}{20}$. 2. $\frac{1}{20}$. 3. $\frac{15}{56}$. 4. $\frac{2}{99}$. 5. $\frac{1}{24}$. 6. $\frac{3}{5}$.
7. $\frac{11}{35}$. 8. $\frac{4}{65}$. 9. $\frac{5}{6}$. 10. $\frac{7}{10}$. 11. $1\frac{3}{4}$. 12. 4.
13. $1\frac{1}{5}$. 14. $\frac{4}{5}$. 15. $\frac{3}{14}$. 16. 8. 17. $\frac{2}{3}$. 18. $\frac{7}{12}$.
19. $2\frac{5}{12}$. 20. $1\frac{23}{24}$. 21. $\frac{1}{10}$. 22. $1\frac{3}{5}$. 23. 40. 24. $\frac{2}{3}$.
25. (a) $\frac{7}{24}$. (b) $\frac{1}{12}$. (c) $\frac{3}{32}$. (d) $\frac{2}{15}$.
26. (a) $\frac{5}{32}$. (b) $\frac{5}{16}$. (c) $\frac{1}{16}$. (d) $\frac{1}{32}$.
27. (a) $\frac{17}{156}$. (b) $\frac{121}{156}$.
28. (a) 0. (b) $1\frac{4}{91}$. (c) $1\frac{3}{8}$. (d) $3\frac{3}{4}$.
29. (a) 57. (b) 62. (c) 98. (d) 3.
30. (a) $\frac{4}{17}$. (b) $\frac{6}{11}$. (c) $\frac{2}{3}$. (d) $\frac{7}{16}$.
31. 437·5 m. 32. \$5·60. 33. 19·6 m.
34. £1260. 35. £15. 36. 39 kg.
37. £364·50. 38. 28p A. 70p B. 98p C.
39. £60. 40. £67·50.
41. £4·55 (a) £3·60 (b). 42. 1 kg.

Exercise 3

1. 2·8. 2. 3·3. 3. 48·6.
4. 3. 5. 0·24. 6. 4·5.
7. 0·8. 8. 20. 9. 0·2.
10. 80. 11. 0·005. 12. 900.
13. 1984·692. 14. 382·226. 15. 203·5944.
16. 91·462. 17. 4·527. 18. 0·0342.
19. $1\frac{1}{3}$ (a). 5 (b). 1 (c). 20. 316·25.
21. 75 m. 22. 5 m/s.
23. 120 l. 24. 28 m.
25. 16. 26. 2 kg (a). 2 t (b). 500 m (c).
27. £4. 28. £2.
29. £28·50. 30. 0·25 (a). 0·2 (b).

Exercise 4

1. 3·461 kg (a). 0·056 427 kg (b).
2. 475 000 cm² (a). 3·65 cm) (b).
3. 54 l. 4. 14·4 km/h. 5. 27·5 gal.
6. 55·1 tons. 7. $56\frac{1}{4}$ miles. 8. 152·727 kg.
9. 1 h 20 min. 10. 22 hl. 11. 202·43 ha.
12. 15 miles/h. 13. £193·60. 14. 0·4049 ha.

15. 810 f/s. 16. 67½ miles/h. 17. 10·87 miles/h.
18. 64 km/h. 19. £299·64. 20. 34·72.
21. (*a*) 48 km/h; 64 km/h; 112 km/h.
    (*b*) 50 km/h; 60 km/h; 110 km/h.

Exercise 5

1. 49·03. 2. 542·13. 3. 956·349.
4. 1898·373. 5. 161·8. 6. 5086·83.
7. 5051. 8. 200 175 000. 9. 148.
10. 243 000 000. 11. 85·0. 12. 18 500.
13. 0·16. 14. 20. 15. 19.
16. 22·3. 17. 22·27. 18. 0·624.
19. 0·89. 20. 0·0689. 21. 3645.
22. 74·8 m. 23. 458·7 m. 24. 83p.

Exercise 6

1. 95p. 2. £2·40. 3. £2·56.
4. £4·20. 5. £2·10. 6. £8·40.
7. £7·80. 8. £12. 9. £5·10.
10. £11·60. 11. £1·35. 12. £4·95.
13. 20%. 14. 9%. 15. 7½%.
16. 25%. 17. 12½%. 18. 20%.
19. 8%. 20. 10%. 21. 75%.
22. 37½%. 23. 80%. 24. 66⅔%.
25. 71³⁄₇%. 26. $\frac{3}{20}$. 27. $\frac{1}{8}$.
28. $\frac{9}{20}$. 29. $\frac{3}{10}$. 30. $\frac{5}{8}$.
31. £18 275. 32. 78% (*a*). 22% (*b*).
33. £6750. 34. £2·96. 35. 15%.
36. 13·97%. 37. 4·9%. 38. £7080.
39. £23·37. 30. £84. 41. £8·55.
42. £60. 43. £6. 44. 93 000.

Exercise 7

1. £15·74. 2. £1556·40. 3. £27·62.
4. £4600. 5. £2037.

Exercise 8

1. 25% (*a*). 20% (*b*). 2. 31⁷⁄₁₀% (*a*). 24¹⁄₁₀% (*b*).
3. 10%. 4. 30·4% (*a*). 9·87% (*b*).

5. £159 500.   6. £13 162·50 (a).   £36 562·50 (b).
7. £1178.   8. £2167·50.   9. £500.
10. 35p.   11. £75.   12. £50·40.
13. 5%.   14. £225·60.   15. £9·90.
16. 50p.   17. 4½% gain.   18. £10·50.
19. £4.   20. 66⅔% (a). 40% (b).

Exercise 9

1. £3·68.   2. £16·20.   3. £15·30.
4. £27·00.   5. £175·50.   6. £5·94.
7. £24·50.   8. £37·12.   9. £31·73.
10. £14·65.   11. £292·60.   12. £669·02.
13. £839·38.   14. £394·13.   15. £892·20.
16. £150.   17. £133·33.   18. £360.
19. £186·67.   20. £320.   21. 12%.
22. 7½%.   23. 10%.   24. 4½%.
25. 8½%.   26. 7 months.   27. 146 days.
28. 4 months.   29. 200 days.   30. 10 months.
31. £16·77.   32. £58·14.   33. £63·06.
34. £43·20.   35. £31·71.   36. £508·40.
37. 25th July.   38. £800.   39. £7·54.
40. 9 yr. 6 months.   41. £13·86.   42. £25 238·36.

Exercise 10

1. £516·91.   2. £740·88.   3. £865·70.
4. £620·88.   5. £484·38.   6. £228·23.
7. £44·60.   8. £69·58.   9. £27·99.
10. £112·38.   11. £39·35.   12. £129·30.
13. £15·76.   14. £518·30.
15. 20·40 (a).   £20·61 (b).   16. £60 480.
17. £43·93.   18. £2·19.   19. £512·90.
20. £535·94.   21. 8·65%.   22. £7382·71.
23. £458·61.   24. £30 593.

Exercise 11

1. 100 m; 160 m; 180 m.   2. 10:11.
3. 9:10.   4. 24:25.   5. 84 kg.
6. 31:56.   7. £323·64.   8. 121·3 km.

9. 24 tonnes.   10. 4 h.   11. £42; £48; £50.
12. 42 kg.   13. £29·12.   14. 12 ha.
15. 12 days.   16. 45p.
17. 480 (*a*). 60 dm³ (*b*). 4 kg/cm² (*c*).   18. 5 men.
19. £26·14 (*a*). £99·34 (*b*). £73·20 (*c*).   20. 30 h.

Exercise 12

1. £3·20.   2. £153.
3. 40p. A.   60p. B.   £1·40. C.   4. £9.
5. £1000.   6. £9000.
7. £1125. A.   £1875. B.   £1250. C.   8. 7 : 3
9. £3750. A.   £6750. B.
10. £525. A.   £700. B.   £210. C.
11. £930. A.   £880. B.   £730. C.
12. £2000. A.   £1500. B.
13. £2691·66. A.   £901·67. B.   £906·67. C.
14. £618. A.   £412. B.   £206. C.
15. £1330. A.   £1520. B.

Exercise 13

2. 11.   3. 16.   4. 1·5.   5. 17.
6. 0·11.   7. 1·2.   8. 0·16.   9. 90.
10. $2^3 \times 3^3$.   11. $5^2 \times 11$.   12. $2^6 \times 3^3$.
13. $3^2 \times 5^3$.   14. $2^5 \times 3^2 \times 5 \times 11^2$.
15. 72.   2520.   16. 72.   10 296.   17. 225.   4725.
18. 144.   2160.   19. 1008.   144 144.   20. 27.
21. 57.   22. 17.   23. 11.
24. £9·24.   25. 10p.   26. 6 g.
27. 289·4.   28. 487·7.   29. 0·2944.
30. 1·682.   31. $2^2 \times 3^4$ (*a*).   18 (*b*).
32. $2^2 \times 3^2 \times 11^2$ (*a*).   66 (*b*).
33. $2^4 \times 3^4$ (*a*).   36 (*b*).   34. $2^4 \times 11^2$ (*a*).   44 (*b*).
35. $7^2 \times 11^2$ (*a*).   77 (*b*).   36. $2^2 \times 3^2 \times 7^2$ (*a*).   42 (*b*).
37. 16 weeks.   38. 221.

Exercise 14

1. 52.   2. 26·62 ha.   3. 224 m.
4. 1136 m.   5. 5·4 dm³.   6. 20 m.
7. 15 dm²   8. £1·61   9. £35·70.

10. £20.   11. £7776.   12. 183·75 kg.
13. 22·5 m.   14. 4 mm.   15. 800 g.
16. £45.   17. £172·50.
18. 1·75 m (*a*). 119 (*b*).   19. £10·50.
20. £58·80.   21. 5·04 tonnes.

Exercise 15

1. 70 cm².   2. 70 m.   3. 1176 m².
4. 14·7 cm².   5. 15 cm.   6. 6 cm.
7. 17 cm.   8. 20 cm³.   9. 66 m.
10. 447 m.   11. 2·1 m.   12. 200 m².
13. 37·206 kg.   14. 6·4 m.   15. 45·7 cm².
16. 7960 m².

Exercise 16

1. 33 cm.   2. 24·5 cm.   3. £55.
4. 1·76 m.   5. 20 cm².   6. 9·63 m².
7. 49·6 m.   8. 379.   9. 132 cm².
10. 422 g.   11. £61·60.   12. 14·08 m².
13. 59·4 kg.   14. 8 min 29 s.   15. 125·714 kg.
16. 10 cm.   17. 38 cm.

Exercise 17

1. £4·54.   2. 11 392·5 fr.   3. $455.
4. £1600.   5. 16 000 rupees.   6. £320.
7. £83·64.   8. 51·1 fr. gain.   9. £11·02.
10. £48.   11. 6000 pes.   12. 172·80 pes.
13. £300.   14. 860·40 kr.   15. £263·89.

Exercise 18

1. 17p.   2. 17·2 fr.   3. $1·83.
4. £1·86.   5. £34·96.   6. 716 pes.
7. C by 2 m.   8. 5·938.   9. 11p per lb.
10. 79c. per kg.   11. £1534.   12. $17·98.
13. 455·34 fr.   14. $1·8.   15. 4·27 fr.
16. $980.

Exercise 19

1. £1·24.   2. £95·48.   3. 21·4p.
4. £1140·48.   5. £28.   6. £31·2.

7. £99. 8. £2 640 15o. 9. £562·50.
10. £331·50. 11. £804. 12. £1401·75.
13. (a) 159. (b) 236. 14. £17 095·50 (a). £418·50 (b).
15. £5025·8o 16. £29·90. 17. £546·25.

Exercise 20

1. 70. 2. 1·65 m. 3. 6p.
4. £45·6o. 5. 40 years. 6. 8·6 km/h.
7. £76·77. 8. £2006. 9. £2·31.
10. 5:3 11. £2·34½.
12. 750 kg. 1 t. 1·25 t. 13. £200 in x. £400 in y.
14. £600 in x. £200 in y. 15. £426.

Exercise 21

1. £1140. 2. £77·70. 3. £27.
4. £7230. 5. 12 000. 6. £52 000.
7. £48 100. 8. £31 320. 9. £278·24.
10. £617·28. 11. £116·64. 12. £322·17.
13. £813·74. 14. 4·9%. 15. £21 628·10.
16. £14. 17. £910. 18. 42½p.
19. 77½p. 20. £12 027.

Exercise 22

1. £288·65. 2. £354·67. 3. £9·25.
4. £17·55. 5. £470·64. 6. £622·01.
7. £344·26. 8. £118·81. 9. 16p.
10. 44p.

Exercise 23

1. 84 km/h. 2. 175 miles/h. 3. 6 miles/h.
4. 600 km/h. 5. 67 m/s.
6. 120 km/h (a). 33 m/s (b). 7. 50 s.
8. 126·72 km/h. 9. 59·1 km/h.
10. 1·822 km from house. 9.40 p.m. 11. 22·2 km/h.
12. 52½ min past 11.0 a.m. 18·75 km from A.
13. 8 km/h. 14. 80 km/h. 15. 19·5 km/h.
16. 20 km/h.

Exercise 24

2. 12·5 l (*a*).   26·4 pt (*b*).   3. £23·40 (*a*).   36 h (*b*).
4. 38·4 km (*a*).   22½ miles (*b*).   5. 11 km/h.
7. 12 720 A.   9540 B.   15 900 C.

Conversion of Common British Units to Equivalent Values in SI Units

Length

1 in	=	25·4 mm = 2·54 cm
1 ft	=	0·3048 m
1 yd	=	0·9144 m
1 mile	=	1·609 34 km

Area

1 in² (square inch)	=	645·16 mm² = 6·4516 cm²
1 ft² (square foot)	=	0·0929 m² = 929 cm²
1 yd² (square yard)	=	0·8361 m²
1 acre	=	4047 m² = 0·4047 ha
1 sq mile	=	2·59 km² = 259 ha

Volume

1 in³ (cubic inch)	=	16·387 cm³
1 ft³ (cubic foot)	=	0·0283 m³ = 28·3 dm³
1 yd³ (cubic yard)	=	0·765 m³

Capacity

1 pt (pint)	=	0·568 dm³ = 0·568 litre
1 qt (quart)	=	1·137 dm³ = 1·137 litres
1 gal	=	4·546 dm³ = 4·546 litres

Mass

1 gr (grain)	=	64·8 mg
1 dr (dram)	=	1·77 g
1 oz	=	28·35 g
1 lb	=	0·4536 kg
1 cwt	=	50·8 kg
1 ton	=	1016·05 kg
1 oz troy	=	31·1035 g
1 dwt (pennyweight)	=	1·555 g

Conversion of Common SI Units to Equivalent Values in British Units

Length

1 km	=	0·621 371 mile
1 m	=	1·093 61 yd
1 mm	=	0·039 3701 in

Area

1 km²	=	247·105 acres
1 m²	=	1·195 99 yd²
1 mm²	=	0·001 55 in²

Volume

1 m³	=	1·307 95 yd
1 dm³	=	0·035 314 7 ft³
1 cm³	=	0·061 023 7 in³
1 l (litre)	=	0·220 gal

Mass

1 kg	=	2·204 62 lb
1 g	=	0·035 274 oz
	=	15·4324 gr (grain)

Imperial Weights and Measures

Measures of Length

12 inches	= 1 foot.
3 ft	= 1 yard.
6 ft	= 1 fathom.
22 yd	= 1 chain.
10 chains	= 1 furlong.
8 fur, or 1760 yd, or 5280 ft	= 1 mile.
6080 ft	= 1 nautical mile (a speed of 1 nautical mile per hour = 1 knot).

Square Measure

144 in²	= 1 square foot.
9 ft²	= 1 square yard.
1210 yd²	= 1 rood.
4 roods, or 4840 yd²	= 1 acre.
640 acres	= 1 square mile.

Cubic Measure

1728 in³	= 1 cubic foot.
27 ft³	= 1 cubic yard.

Measures of Weight

Avoirdupois Weight

16 drams	= 1 ounce.
16 oz, or 7000 grains	= 1 pound.
14 lb	= 1 stone.
28 lb	= 1 quarter.
4 qr, or 112 lb	= 1 hundredweight.
20 cwt, or 2240 lb	= 1 ton.
2000 lb	= 1 short ton.

Troy Weight

24 grains	= 1 pennyweight.
20 dwt	= 1 ounce.
12 oz, or 5760 grains	= 1 pound (seldom used).

Measures of Capacity

4 gills	= 1 pint.
2 pt	= 1 quart.
4 qt	= 1 gallon.
2 gal	= 1 peck.
4 pk	= 1 bushel.
8 bush	= 1 quarter.